Electronic Warfare Target Location Methods

For a listing of recent titles in the *Artech House Information Warfare Library*,
turn to the back of this book.

Electronic Warfare Target Location Methods

Richard A. Poisel

ARTECH HOUSE

BOSTON | LONDON
artechhouse.com

Library of Congress Cataloging-in-Publication Data
Poisel, Richard A.
A catalog record for this book is available from the Library of Congress.

British Library Cataloguing in Publication Data
Poisel, Richard
Electronic warfare target location methods. —(Artech House electronic warfare library)
1. Target acquisition—Mathematical models 2. Electronics in military engineering
3. Algorithms
I. Title
623'.043
ISBN 10: 1-58053-968-8
ISBN 13: 978-1-580-53968-5

Cover design by Yekaterina Ratner

685 Canton Street
Norwood, MA 02062

International Standard Book Number: 1-58053-968-8

10 9 8 7 6 5 4 3

To my wife, Debbie

Table of Contents

Preface

This book was written for the purpose of presenting in a single place many of the algorithms used for computing the geolocations of targets. Determining the geographical coordinates of targets is one of the fundamental functions of electronic warfare (EW) systems.

Geoposition determination is a much studied topic. Ancient mariners used the stars to determine their position while at sea. Currently, satellite systems such as the Global Positioning System serve the same purpose. For years, systems such as LORAN and OMEGA were used for the both air and marine navigation.

For the EW application of geopositioning, the fundamental difference is one of cooperation. The target system is not cooperating with the EW system when the latter is trying to locate the former. This issue alone significantly complicates the attempts at geopositioning. For one, the times of the target transmissions are not known. The EW system must find in frequency-time where the correct transmission lies. These issues are not covered here; some of those issues are covered in a related book [1].

Chapter 1 provides an introduction to the topic of target geolocation and presents some terminology. Also presented in Chapter 1 is one of the fundamental algorithms used for computing position fixes (PF), the gradient descent method. It is presented in the first chapter because it does not depend on any particular type of parameter measurement. In fact, mixed parameters can be combined in the algorithm.

Chapter 2 presents algorithms for PF estimation based on triangulation. This is where the intersection of two or more lines of bearing (LOB) is determined. At that intersection, or in the region around the intersection when there are more than two LOBs, is where the PF computation estimates the target to be. There are several ways to determine this estimate.

PF estimation based on measuring time parameters is presented in Chapter 3. Using the times of arrival, time differences of arrival, or range differences produces quadratic lines of position (LOP), which may be any of the standard conics or not-so-standard second-order functions. Nevertheless, again it is the intersection of these LOPs that produces the PF estimate. History has shown that these techniques are generally more accurate than triangulation.

High frequency (HF) signals can propagate over considerable distances. This usually occurs in the lower third of the HF band. The propagation phenomena that allow for this are refraction of the signal by the ionosphere, which bends the electromagnetic (EM) wave, causing its propagation direction to head back toward the Earth, and reflection of signals by the Earth's surface. This reflection sends the EM wave back toward the ionosphere. This refraction/reflection sequence can occur several times, facilitating signal propagation over considerable distances.

When there is only one such hop, however, the geolocation of the transmitter can be estimated with a properly configured single-sensor system by measuring the azimuth angle of arrival (AOA) and the elevation AOA. Knowledge of the effective height of the ionosphere at the equivalent point of reflection is required. Such methods carry the appellation single-site location (SSL). SSL techniques are discussed in Chapter 4.

All of the methods of obtaining the raw parameters used for PF estimation are subject to measurement errors and noise. Therefore, it is rare that multiple LOBs or LOPs all intersect at the same point. The PF estimation methods attempt to deal with this uncertainty in different ways, some better than others. Another source of error not due to uncertainty in the basic measurements is the limited sample size. Some estimation methods are asymptotically unbiased, while others are biased. The asymptote in this case is in the limit as the number of samples approaches infinity. With few samples, which is normally the case, computation errors can arise, causing erroneous estimates. Error sources and methods to deal with them are included.

This book is intended as an introduction to the subject of target location estimation. The target audiences are engineers and scientists with a minimum of a four-year technical degree. While this book may be of use to practicing engineers as a reference, the principal reader is intended to be one new to the field of electronic warfare.

Reference

[1] Poisel, R.A., *Target Acquisition in Electronic Warfare Systems,* Norwood, MA: Artech House, 2004.

Chapter 1

Introduction to Emitter Geolocation

Determining the location of an emitting target is one of the fundamental functions of communication *electronic warfare* (EW) systems. Knowing the location of targets is useful for several purposes. First, knowing the location of targets indicates the disposition of forces. Second, precision location of targets allows for use of *global positioning system* (GPS)-enabled fire-and-forget munitions for negation of the target. Third, clustering different types of emitters in the same region can give an indication of the type of entity at a particular location.

This monograph presents several modern (and some not-so-modern) techniques for computing the *position fix* (PF) of a target based on differing information available from measurements performed on intercepted signals.

The azimuth angle of arrival of a signal, or its *line of bearing* (LOB), is a frequently used parameter for PF computation. Two or more LOBs, assumed to be measured on the same target at more or less the same time, may intersect as illustrated in Figure 1.1. Such a technique for PF determination is referred to as *triangulation.* Triangulation is discussed at length in Chapter 2, where several different algorithms are presented.

Another possibility is to measure the *time of arrival* (TOA) of the signal at several sensors. The TOA itself can be used to compute the PF, where the isocontours are circles, but more typically the TOAs are transferred to a central site where the *time differences of arrival* (TDOA) are computed between the sensors, two at a time. Closely related to the method of using the TDOAs is to compute the range differences between the sensors and the target. Such *range differences* (RD) are related to the TDOAs by the speed of propagation in the medium through which the signal propagates. In the air this is normally assumed to be the speed of light. If the signal is an audio signal, then in the air it is the speed of propagation of sound through the air, which can be influenced by humidity and other parameters in the air. If the signal is an audio signal underwater, then the speed of propagation is that of sound through water.

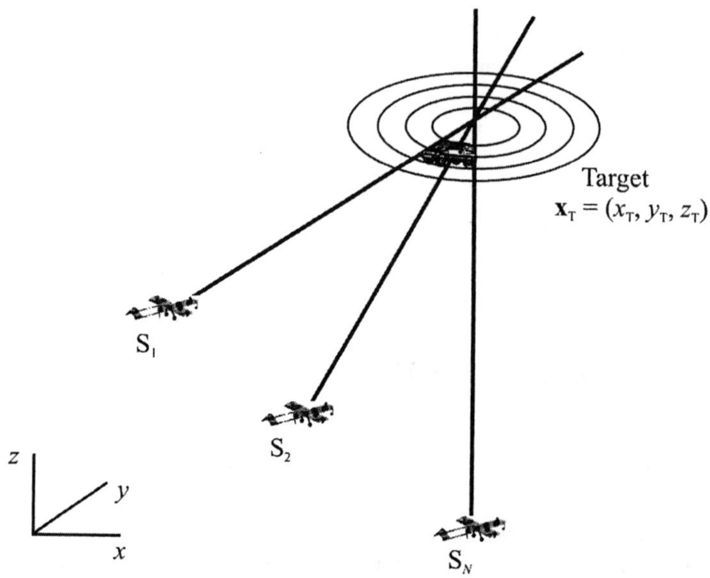

Figure 1.1 Intersection of measured LOBs.

All of the techniques that use the TOA, TDOA, or RD generate quadratic curves, called *lines of position* (LOP), upon which the emitter lies, subject to measurement errors and noise perturbations. The intersection of these LOPs is used to estimate the PF. Quadratic PF techniques are discussed in Chapter 3. As discussed in Chapter 3, if the target is moving, errors can occur in the PF calculations. These errors can be mitigated if the motion is detected.

Signals in the *high frequency* (HF) range can propagate for considerable distances using reflections off the Earth's ionosphere. Such bending of the signal is actually being caused by refraction of the signals within the ionosphere due to variations in the electron and ion density in the ionosphere. Knowing the azimuth angle of arrival of the signal, as well as its elevation angle, combined with knowledge of the height of the ionosphere allows estimation of the PF of a target with a single sensor. Such PF techniques are known as *single-site location* (SSL). SSL methods are discussed in Chapter 4.

In general, the methods of measurement of the necessary parameters are not covered here—they are covered in several other places [1–6]. The exceptions to this are in Chapter 3, where the cepstrum is discussed for computing time parameters and techniques are discussed for determining the TDOA and differential doppler (DD) with phase data.

Table 1.1 Advantages and Disadvantages of the Gradient Descent Approach to PF

Advantages	Disadvantages
Multiple independent measurements to a single station are averaged naturally	The method is iterative, requiring an initial guess
Multiple measurements and mixed-mode measurements are combined properly, that is, with the correct geometric factors, and can be weighted according to their a priori accuracies	It is computationally complex compared to simple plotting of lines of position
The statistical spread of the solution can be found easily and naturally	Being a local correction, its convergence is not assured
Experience indicates that the initial position guess can be quite far off without preventing good convergence	
Failure to converge is easy to detect	
Simulation is easy, so convergence can be readily tested	
Computational complexity is less than that of a Kalman-Bucy filter	

Source: [7].

1.1 Gradient Descent Algorithm

Foy presented a PF algorithm based on the simple process of using a Taylor-series expansion of the defining equations (also known as Gauss or Gauss-Newton interpolation) [7]. This development provides a convenient introduction to the algorithmic approach to computing PFs. It contains many of the characteristics typical of solving the PF estimation problem, so it is presented in this introduction. It is also one of the most general and most accurate methods available. Other types of descent algorithms exist and can be used as well [6].

The equations express the geometry involved in mathematical terms and are frequently nonlinear. Expanding the nonlinearities in a Taylor series and retaining only the linear terms allows for employing Newton-Raphson methods of gradient descent to iteratively find a solution. An initial guess of the solution is made, and the estimation process proceeds iteratively until an estimated solution is found. At each stage, a correction factor is calculated based on the local linear least-sum square error.

The principal disadvantages and advantages according to Foy are as shown in Table 1.1.

Suppose $\mathbf{x}_\text{T} = (x_\text{T}, y_\text{T})$ represents the true position of the target emitter and $\mathbf{x}_k = (x_k, y_k)$, $k = 1, 2, \ldots, N$ represents the true positions of the N PF sensors.[1, 2] Let m_{ki} represent the ith measurement at sensor k. These measurements, at this point,

[1] Bold capital letters refer to matrices, small bold letters refer to vectors, and italicized letters refer to variables.

[2] This analysis is restricted to two dimensions. Extension to higher dimensionality is straightforward.

are not restricted to LOBs, ranges, and so forth, but could represent any measurement useful for calculating the PF. Thus,

$$f_i(x_T, y_T, x_k, y_k) = u_i = m_{ki} + \epsilon_i, \quad k = 1, 2, \cdots, N \tag{1.1}$$

where

m_{ki} = true value of the measured quantity

and

ϵ_i = error in the m_{ki} measurement

For example, if the measurement is that of an LOB from the PF sensor to the emitter, then

$$f_i(x_T, y_T, x_k, y_k) = \tan^{-1}\left(\frac{y_T - y_k}{x_T - x_k}\right) + \epsilon_i \tag{1.2}$$

The goal is to find (x_T, y_T) given the measurements and known locations of the sensors. The errors in the measurements ϵ_i are assumed to be *independent* (of each other) *and identically distributed* (i.i.d.) with zero means, $\mathcal{E}\{\epsilon_i\} = 0$, where $\mathcal{E}\{\}$ denotes statistical expectation.[3] The *error covariance matrix* is given by

$$R = [r_{ij}] \tag{1.3}$$

with entries

$$r_{ij} = \mathcal{E}\{\epsilon_i \epsilon_j\} \tag{1.4}$$

Let $(\hat{x}_T, \hat{y}_T)$ be guesses of the true position (x_T, y_T). Then

$$x_T = \hat{x}_T + \delta_x \tag{1.5}$$

and

$$y_T = \hat{y}_T + \delta_y \tag{1.6}$$

[3] Strictly speaking, specification of expectations requires a domain over which the expectation applies. When that domain is obvious, it will not be indicated.

The Taylor series expansion of $f_i()$ around point a is given by

$$f(x) = f(a) + (x-a)f'(a) + \frac{(x-a)^2}{2!}f''(a) + \cdots + \frac{(x-a)^n}{n!}f^n(a) + \cdots \quad (1.7)$$

where the notation $f^n(a)$ refers to the nth derivative of $f(x)$ evaluated at point a. In two dimensions, this becomes

$$\begin{aligned} f(a+h, b+k) = f(a,b) &+ \left(h\frac{\partial}{\partial x} + k\frac{\partial}{\partial y}\right) f(x,y)\Big|_{\substack{x=a\\y=b}} + \cdots \\ &+ \frac{1}{n!}\left(h\frac{\partial}{\partial x} + k\frac{\partial}{\partial y}\right)^n f(x,y)\Big|_{\substack{x=a\\y=b}} + \cdots \end{aligned} \quad (1.8)$$

where the bar and subscripts mean that, after differentiation, x is replaced with a and y is replaced with b. Also, in this notation,

$$\left(h\frac{\partial}{\partial x} + k\frac{\partial}{\partial y}\right) f(x,y) = \left(h\frac{\partial f(x,y)}{\partial x} + k\frac{\partial f(x,y)}{\partial y}\right)$$

$$\left(h\frac{\partial}{\partial x} + k\frac{\partial}{\partial y}\right)^2 f(x,y) = h^2\frac{\partial^2 f(x,y)}{\partial x} + 2hk\frac{\partial^2 f(x,y)}{\partial x \partial y} + k^2\frac{\partial f(x,y)}{\partial y}$$

$$\cdots$$

For the case at hand, the two-dimensional Taylor expansion, after deleting all terms higher than the linear ones, is

$$\hat{f}_i + \frac{\partial f_i()}{\partial x}\Big|_{\substack{x=\hat{x}_T\\y=\hat{y}_T}} \delta_x + \frac{\partial f_i()}{\partial y}\Big|_{\substack{x=\hat{x}_T\\y=\hat{y}_T}} \delta_y \approx m_{ki} + \epsilon_i \quad (1.9)$$

where

$$\hat{f}_i = f_i(\hat{x}_T, \hat{y}_T, x_k, y_k) \quad (1.10)$$

To put this development into matrix form for easier manipulation, define

$$\mathbf{H}=\begin{bmatrix} h_{11} & h_{12} \\ h_{21} & h_{22} \\ \vdots & \vdots \\ h_{N_S1} & h_{N_S2} \end{bmatrix}=\begin{bmatrix} \left.\dfrac{\partial f_1()}{\partial x}\right|_{\substack{x=\hat{x}_T\\ y=\hat{y}_T}} & \left.\dfrac{\partial f_1()}{\partial y}\right|_{\substack{x=\hat{x}_T\\ y=\hat{y}_T}} \\ \left.\dfrac{\partial f_2()}{\partial x}\right|_{\substack{x=\hat{x}_T\\ y=\hat{y}_T}} & \left.\dfrac{\partial f_2()}{\partial y}\right|_{\substack{x=\hat{x}_T\\ y=\hat{y}_T}} \\ \vdots & \vdots \\ \left.\dfrac{\partial f_{N_S}()}{\partial x}\right|_{\substack{x=\hat{x}_T\\ y=\hat{y}_T}} & \left.\dfrac{\partial f_{N_S}()}{\partial y}\right|_{\substack{x=\hat{x}_T\\ y=\hat{y}_T}} \end{bmatrix} \tag{1.11}$$

$$\boldsymbol{\delta}=\begin{bmatrix} \delta_x \\ \delta_y \end{bmatrix} \tag{1.12}$$

$$\mathbf{z}=\begin{bmatrix} m_{k1}-\hat{f}_1 \\ m_{k2}-\hat{f}_2 \\ \vdots \\ m_{kN_S}-\hat{f}_{N_S} \end{bmatrix} \tag{1.13}$$

and

$$\mathbf{e}=\begin{bmatrix} \epsilon_1 \\ \epsilon_2 \\ \vdots \\ \epsilon_{N_S} \end{bmatrix} \tag{1.14}$$

then (1.9) can be written

$$\mathbf{H}\boldsymbol{\delta}\approx\mathbf{z}+\mathbf{e} \tag{1.15}$$

The form of $\boldsymbol{\delta}$ that yields the least sum squared error with the terms weighted according to the covariance matrix is [8]

$$\boldsymbol{\delta}=[\mathbf{H}^T\mathbf{R}^{-1}\mathbf{H}]^{-1}\mathbf{R}^{-1}\mathbf{z} \tag{1.16}$$

where $\mathbf{R}$ is a weighting matrix used to factor in one or more parameters. $\mathbf{R}$ is arbitrary, but must be positive definite and full rank so that $\mathbf{R}^{-1}$ exists. Therefore, during one step of the iteration, $\boldsymbol{\delta}$ is computed according to (1.16), and new estimates (guesses) are obtained via

$$\begin{aligned} \hat{x}_{\text{new}} &\leftarrow \hat{x}_{\text{old}} + \delta_x \\ \hat{y}_{\text{new}} &\leftarrow \hat{y}_{\text{old}} + \delta_y \end{aligned} \tag{1.17}$$

in (1.15), and the iteration is repeated until there is a satisfactory minimization of change in (x_g, y_g) from one iteration to the next ($\boldsymbol{\delta} \approx \mathbf{0}$).

The covariance matrix of the PF estimate is given by

$$\mathbf{Q}_0 = [\mathbf{H}^T \mathbf{R}^{-1} \mathbf{H}]^{-1} = \begin{bmatrix} \sigma_x^2 & \rho_{xy} \\ \rho_{xy} & \sigma_y^2 \end{bmatrix} \tag{1.18}$$

If the error statistics are normal, then the error region is an ellipse, called the *elliptical error probable* (EEP), with semimajor axis a and semiminor axis b given by [9]

$$a^2 = 2\frac{\sigma_x^2 \sigma_y^2 - \rho_{xy}^2}{\sigma_x^2 + \sigma_y^2 - [\sigma_x^2 - \sigma_y^2 + 4\rho_{xy}^2]^{1/2}} C^2 \tag{1.19}$$

$$b^2 = 2\frac{\sigma_x^2 \sigma_y^2 - \rho_{xy}^2}{\sigma_x^2 + \sigma_y^2 + [\sigma_x^2 - \sigma_y^2 + 4\rho_{xy}^2]^{1/2}} C^2 \tag{1.20}$$

where $C = -2\ln(1 - P_e)$, with P_e being the confidence that the target lies within the EEP (e.g., 0.5 for 50%, 0.9 for 90%, and so forth). The tilt angle θ of the semimajor axis relative to the x-axis is given by

$$\theta = \frac{1}{2}\tan^{-1}\frac{2\rho_{xy}}{\sigma_y^2 - \sigma_x^2} \tag{1.21}$$

The *circular error probable* (CEP) is similar in concept to the EEP. It is a circle centered on the computed PF with an area such that the target lies within the circle with a prescribed probability. It can be estimated to within 10% by [7]

$$\text{CEP} \approx 0.75\sqrt{a^2 + b^2} \tag{1.22}$$

Example: This example is from [7] and is a good illustration of the mixed observation capability of the gradient descent method. Consider two sensors as shown in Figure 1.2. The two sensors are located at (–20, 20) and (8, 12). Three measurements are made: an LOB from sensor S_1 at –38° with $\sigma_\phi = 3^\circ = 0.0524$ radians, a range measurement also from S_1 at $r_1 = 32$ units with $\sigma_r = 2$ units, and a range difference between the two sensors $r_1 - r_2 = \Delta r_{12} = 16$ units with $\sigma_{\Delta_r} = 1$ unit. The bearing can be expressed as

$$\phi_1 = \sin^{-1}\left[\frac{y_1 - y_k}{\sqrt{(x_1 - x_k)^2 + (y_1 - y_k)^2}}\right] \tag{1.23}$$

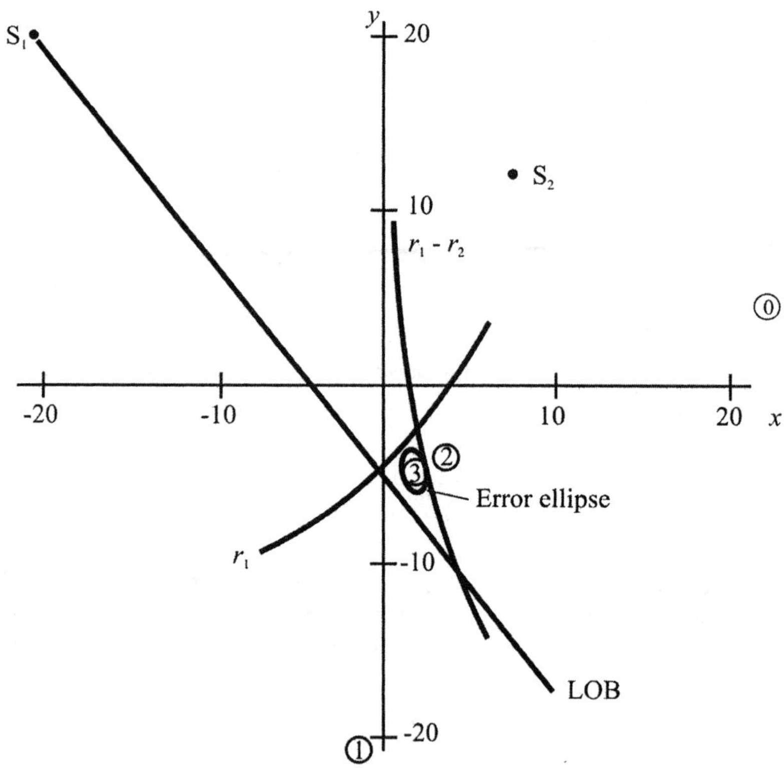

Figure 1.2 Mixed observation example geometry. (*Source:* [7], © 1976 IEEE. Reprinted with permission.)

for which the first-order expansion is

$$\left[\frac{y_1 - y_k}{r_1^2}\right]\delta_x - \left[\frac{x_1 - x_k}{r_1^2}\right]\delta_y \approx \phi_1 - \tan^{-1}\left(\frac{y_1 - y_k}{x_1 - x_k}\right) - e_{\theta_1} \tag{1.24}$$

This result must be weighted by multiplying by the distance between the current guess and S_1, yielding

$$\left[\frac{y_1 - y_k}{r_1}\right]\delta_x - \left[\frac{x_1 - x_k}{r_1}\right]\delta_y \approx r_1\left[\phi_1 - \tan^{-1}\left(\frac{y_1 - y_k}{x_1 - x_k}\right) - e_{\phi_1}\right] \tag{1.25}$$

This is used for a_{11}, a_{12}, and z_1.

The true range from S_1 to the current guess is given by

$$r_1 = \sqrt{(x_k - x_1)^2 + (y_k - y_1)^2} \tag{1.26}$$

The first-order Taylor series expansion is given by

$$\frac{x_k - x_1}{r_{k1}}\delta_x + \frac{y_k - y_1}{r_{k1}}\delta_y \approx r_{\text{measured}} - r_{k1} - e_{r_1} \tag{1.27}$$

while the first-order Taylor series expansion of the range difference is given by

$$\left[\frac{x_k - x_1}{r_{k1}} - \frac{x_k - x_2}{r_{k2}}\right]\delta_x + \left[\frac{y_k - y_1}{r_{k1}} - \frac{y_k - y_2}{r_{k2}}\right]\delta_y \approx \Delta r_{12} - (r_{k1} - r_{k2}) - e_{\Delta r_{12}} \tag{1.28}$$

When the measurement errors are independent, the error covariance matrix of the measurements is

$$\mathbf{R} = \begin{bmatrix} \sigma_\phi^2 & 0 & 0 \\ 0 & \sigma_r^2 & 0 \\ 0 & 0 & \sigma_{\Delta_r}^2 \end{bmatrix} = \begin{bmatrix} (0.0524)^2 r_{1k}^2 & 0 & 0 \\ 0 & 4 & 0 \\ 0 & 0 & 1 \end{bmatrix} \tag{1.29}$$

The algorithm was started with the initial point of $x_0 = 22$ and $y_0 = 4$, as illustrated in Figure 1.2. This results in $\delta_x = -23.5$ and $\delta_y = -24.4$ from (1.16). These values yield $x_1 = -1.5$ and $y_1 = -20.4$ from (1.17). Continuing the algorithm, the final solution at the end of iteration 3 is, as shown in Figure 1.2, $x_3 = 2$, $y_3 = -5.1$, after which $\delta_x = 0.05$ and $\delta_y = -0.05$ and the algorithm halts. At this point,

$$\mathbf{H} = \begin{bmatrix} 0.752 & 0.659 \\ 0.659 & -0.752 \\ 0.990 & 0.192 \end{bmatrix} \tag{1.30}$$

and

$$\mathbf{Q}_0 = \left[\mathbf{H}^{\mathrm{T}}\mathrm{P}^{-1}\mathbf{H}\right]^{-1} = \begin{bmatrix} 0.899 & -0.640 \\ -0.640 & 3.578 \end{bmatrix} \tag{1.31}$$

From (1.19) to (1.21), the EEP parameters are

$$a = 1.929\,\text{units} \qquad b = 0.868\,\text{units} \qquad \theta = 102.76^\circ \tag{1.32}$$

yielding

$$\text{CEP} = 1.587\,\text{units} \tag{1.33}$$

The EEP is plotted in Figure 1.2.

In this example, convergence occurred after only three iterations and the resulting EEP was reasonable, considering the variances of the measurements.

■

1.2 Concluding Remarks

There are many PF estimation methods, some of which are presented in this monograph. There has been no attempt to include all possibilities. GPS alone has spurred considerable interest and research into different techniques of computing PFs (although all of those related to GPS are based on time differences). The ongoing implementation of the Federal Communications Commission (FCC) ruling in the United States mandating geoposition availability of all 911 calls has spurred considerable research into methods for computing PFs as well.

It is hoped that the reader garners a basic understanding of the problems and some useful approaches for estimating the locations of emitters.

References

[1] Jenkins, H. H., *Small-Aperture Radio Direction-Finding,* Norwood, MA: Artech House, 1991.

[2] Poisel, R. A., *Introduction to Communication Electronic Warfare Systems*, Norwood, MA: Artech House, 2002, Chapters 11 and 12.

[3] Neri, F., *Introduction to Electronic Defense Systems,* 2nd Ed., Norwood, MA: Artech House, 2001.

[4] Van Trees, H. L., *Optimum Array Processing, Part IV of Detection, Estimation, and Modulation Theory,* New York: Wiley, 2002.

[5] Cadzow, J. A., "Methods of Direction-of-Arrival," *Proceedings IEEE Pacific Rim Conference on Communications, Computers, and Signal Processing,* May 9–10, 1991.

[6] Cadzow, J. A., "Signal Processing via Least Squares Error Modeling," *IEEE ASSP Magazine,* October 1990, pp. 12–31.

[7] Foy, W. H., "Position-Location Solutions by Taylor-Series Estimation," *IEEE Transactions on Aerospace and Electronic Systems,* Vol. AES-12, No. 2, March 1976, pp. 187–193.

[8] Whalen, A. D., *Detection of Signals in Noise,* New York: Academic Press, 1971, pp. 374–376.

[9] Poisel, R. A., *Introduction to Communication Electronic Warfare Systems*, Norwood, MA: Artech House, 2002, pp. 384–388.

Chapter 2

Triangulation

Triangulation is the appellation applied to estimating a PF by calculating the most likely point for the target, given the intersection of two or more LOBs from sensors at known locations. It is widely used in this role, and there are several techniques available for measuring the azimuth angles of arrival of signals. Noise, measurement errors, and multipath reflections typically limit the PF accuracy.

Triangulation can be implemented on all varieties of platforms, including aircraft, ships, and ground vehicles. If signal phase is used as the parameter for computing the LOBs, then triangulation requires an array of antennas, with a baseline shorter than half a wavelength to avoid ambiguities in phase angle measurement. Parameters other than phase, such as relative amplitude, can also be used as the angle indicator. Most methods, however, require the antenna array.

This chapter starts with a simple illustration of triangulation to introduce the fundamental concepts. Several of the available algorithms for optimum PF calculation are based on the *least-squared error* (LSE) estimation technique. Therefore, a generalized introduction to the LSE method is presented next. An extension to the LSE method is the *total least-squared error* (TLSE) technique, which is somewhat more general and is presented next. PF algorithms based on the method of least-squared distance error are presented. This is followed by a discussion of optimum PF estimation based on the method of minimum mean-square error. A technique based on dividing up the *area of interest* (AOI) into discrete intervals, called the *discrete probability density* (DPD) method, is discussed. The concept of generalized bearings, where the LOBs are not based on the Cartesian coordinates but are three-dimensional, yields a different algorithm for PF estimation; this is discussed next. Using concepts from the statistical process of maximum-likelihood estimation also yields PF estimation algorithms, and two of these are presented. A method of PF estimation based on the intersection of LOB *fans*, defined as ±maximum LOB error, is discussed next. Then discussion shifts to presentations of the theoretical observability requirements for estimation of the PF and velocity of a target. The last topics discussed in this chapter are the sources of error in PF estimation using triangulation. These include systematic errors (bias) and random errors (noise).

2.1 Basic Concepts

The geometry shown in Figure 2.1 can be used to describe the triangulation concept. Initially consider just two dimensions. There are two sensors, S_1 and S_2, separated by distance d, and a single target. Of course, the two sensors could be the same sensor that has moved distance d. Each of the sensors computes an LOB relative to some reference, which is the same for both sensors. By simple trigonometry,

$$\sin\phi_1 = \frac{d_1}{d}$$
$$d_1 = d\sin\phi_1$$

and

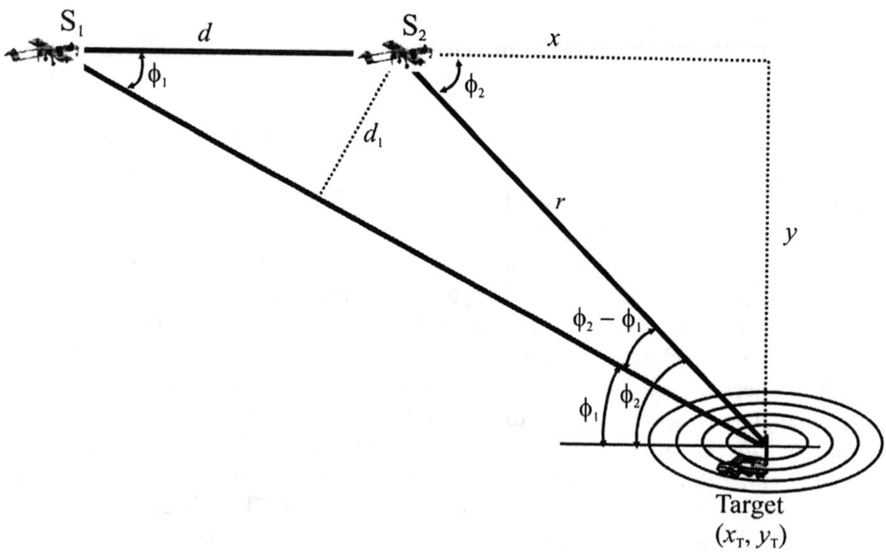

Figure 2.1 Geometric relationships for PF by triangulation.

$$\sin(\phi_2 - \phi_1) = \frac{d_1}{r}$$

$$r = \frac{d_1}{\sin(\phi_2 - \phi_1)}$$

yielding

$$r = \frac{d \sin \phi_1}{\sin(\phi_2 - \phi_1)} \tag{2.1}$$

With knowledge of r and ϕ_2, x and y, the distances to the target from S_2, can be computed as

$$\begin{aligned} x &= r \cos \phi_2 \\ y &= r \sin \phi_2 \end{aligned} \tag{2.2}$$

These notions extend to more than two sensors and to three dimensions in the obvious way, taking two sensors at a time and computing the coordinates of the target. The resulting coordinates can then be averaged. Other methods of using more than two LOBs are discussed in this chapter.

Another technique for PF with triangulation is to plot (figuratively or literally) the measured LOBs and see where they cross, as shown in Figure 2.2 (the sensors could be one moving as shown or three stationary, and the same results would apply). In the absence of errors in the measurement process, these LOBs will all cross at a single point. However, in general, the LOBs are corrupted with measurement error and noise. This noise is frequently assumed to be zero mean *additive white Gaussian noise* (AWGN). The result on the PF computation, however, is to cause the LOBs to move away from crossing at a single point. If the noise is random, the measured LOB could be larger than or smaller than the actual LOB, as illustrated in Figure 2.3. The result is an error ellipse. The sensors could also exhibit biases, which many parameter estimators do. Biases can be caused by being an inherent property of the algorithm to compute the PF or they can be due to systematic errors in the parameter measurement device.

If there are only three LOBs available, then they will form a triangle hopefully close to the actual location of the target, as illustrated in Figure 2.3. There are three popular nonstatistical methods to estimate the location of the target given the triangle. These are shown in Figure 2.4 [1]; they are all techniques for estimating the centroid of the area of the triangle. In Figure 2.4(a), the medians of the sides are connected with the opposite angle. In Figure 2.4(b), the angles are bisected and the point where the bisectors intersect is chosen as the PF. In Figure 2.4(c), the

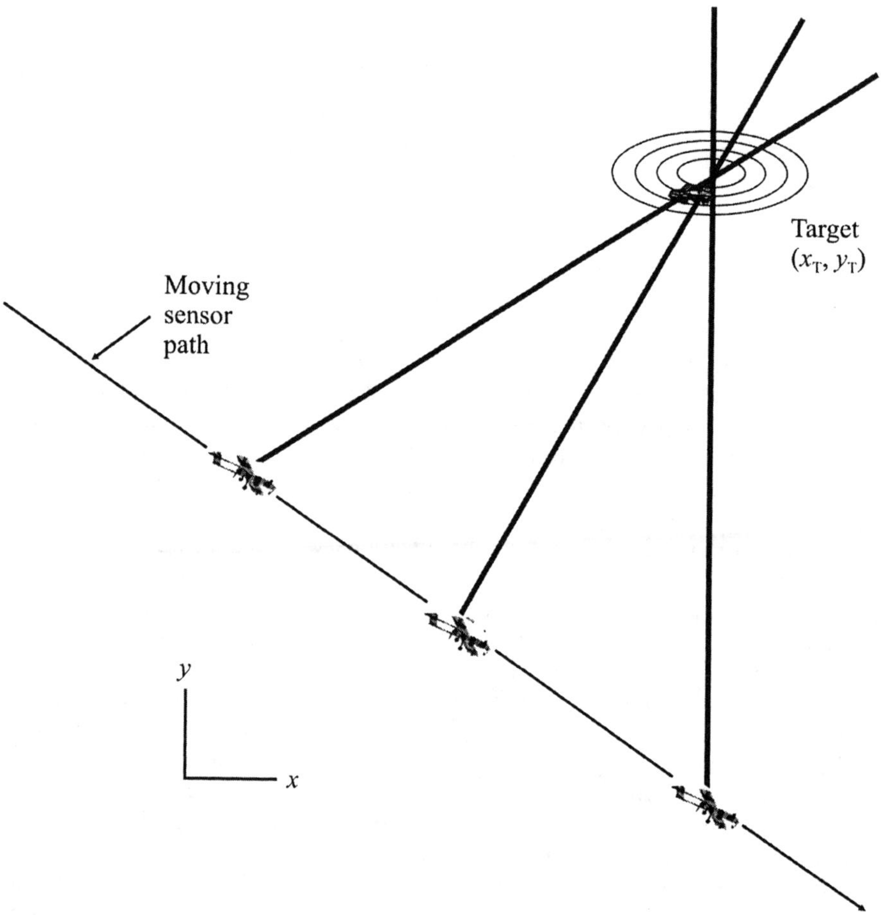

Figure 2.2 Triangulation.

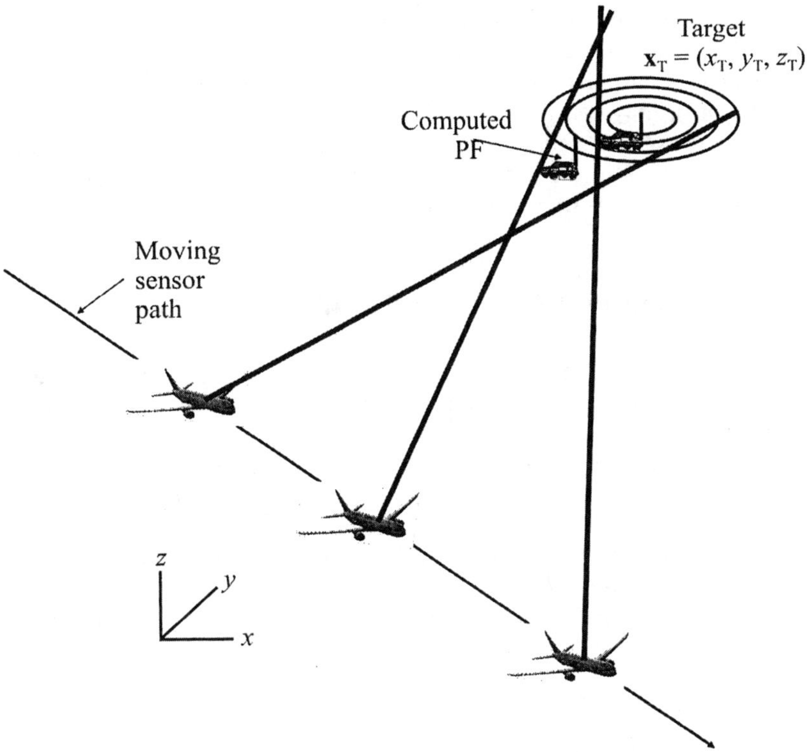

Figure 2.3 Random error effects on computing the PF.

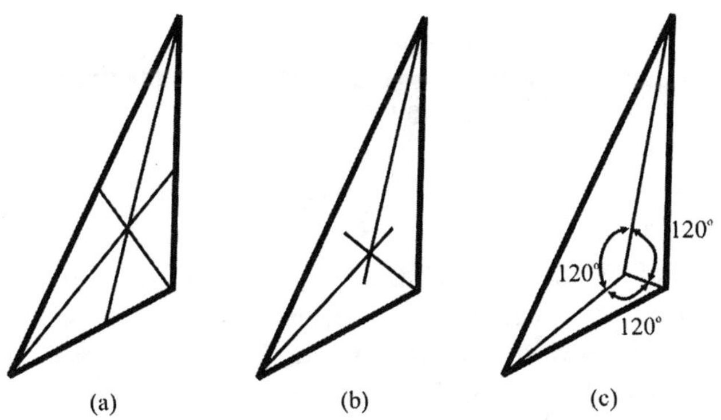

Figure 2.4 Nonstatistical PF calculations with only three LOBs: (a) intersection of medians, (b) intersection of angle bisectors, and (c) Steiner point (defined by the point where the angles between the lines from the corners are all 120° as shown). They are all methods for estimating the centroid of the triangle. (*Source:* [1], © 1991 Horizon House. Reprinted with permission.)

intersection of lines drawn from each angle to the point in the triangle where the lines form 120° angles is chosen as the PF.

If there are more than three LOBs in the calculation, such as those illustrated in Figure 2.5(a), they can be taken three at a time, as shown in Figure 2.5(b–e). The coordinates of the resultant centroids can then be averaged to determine the final fix, as shown by the white circle in Figure 2.6. Note that averaging the intersections of bearings may lead to biased results.

2.2 Least-Squared Error Estimation

Many of the optimization methods to be presented rely on finding an estimate based on minimizing the error between the estimate and the actual value of the PF. Therefore, this section presents the general development of the optimal LSE estimator. Here, the recursive form of LSE estimation is assumed such that measurements are made sequentially at time instants 0, 1, …, k, …, $M-1$. That is, measurements are available at time instants k and before. LSE estimation can be applied to any appropriate set of data points [2], however, to include the case when the estimate is not obtained until all the data points are available.

Let the linear estimation model for the kth time instant be given by

$$\mathbf{x}_k = \mathbf{H}_k \boldsymbol{\theta}_k + \mathbf{n}_k \tag{2.3}$$

where $\mathbf{x}_k$ is the vector of measurements up to time k, and $\mathbf{H}_k$ is the observation matrix such that measurement x_k at time instant k is related to the vector $\boldsymbol{\theta}_k$ by

$$x_k = \mathbf{h}_k \boldsymbol{\theta}_k + n_k \tag{2.4}$$

where $\mathbf{h}_k$ is the kth row vector in $\mathbf{H}_k$ and $\boldsymbol{\theta}_k$ is the unknown parameter vector up to k. Noise term n_k is the measurement noise vector for sample time k.

The estimation model for $\mathbf{x}_k$ is given by

$$\hat{\mathbf{x}}_k = \mathbf{H}_k \hat{\boldsymbol{\theta}}_k \tag{2.5}$$

The goal is to find the optimum estimate for $\boldsymbol{\theta}$, denoted by $\hat{\boldsymbol{\theta}}$, such that some cost function is minimized. For LSE estimation, that cost function is the squared error of the difference between the actual value for $\boldsymbol{\theta}$ and the estimate. Thus, the cost function to be minimized is given by

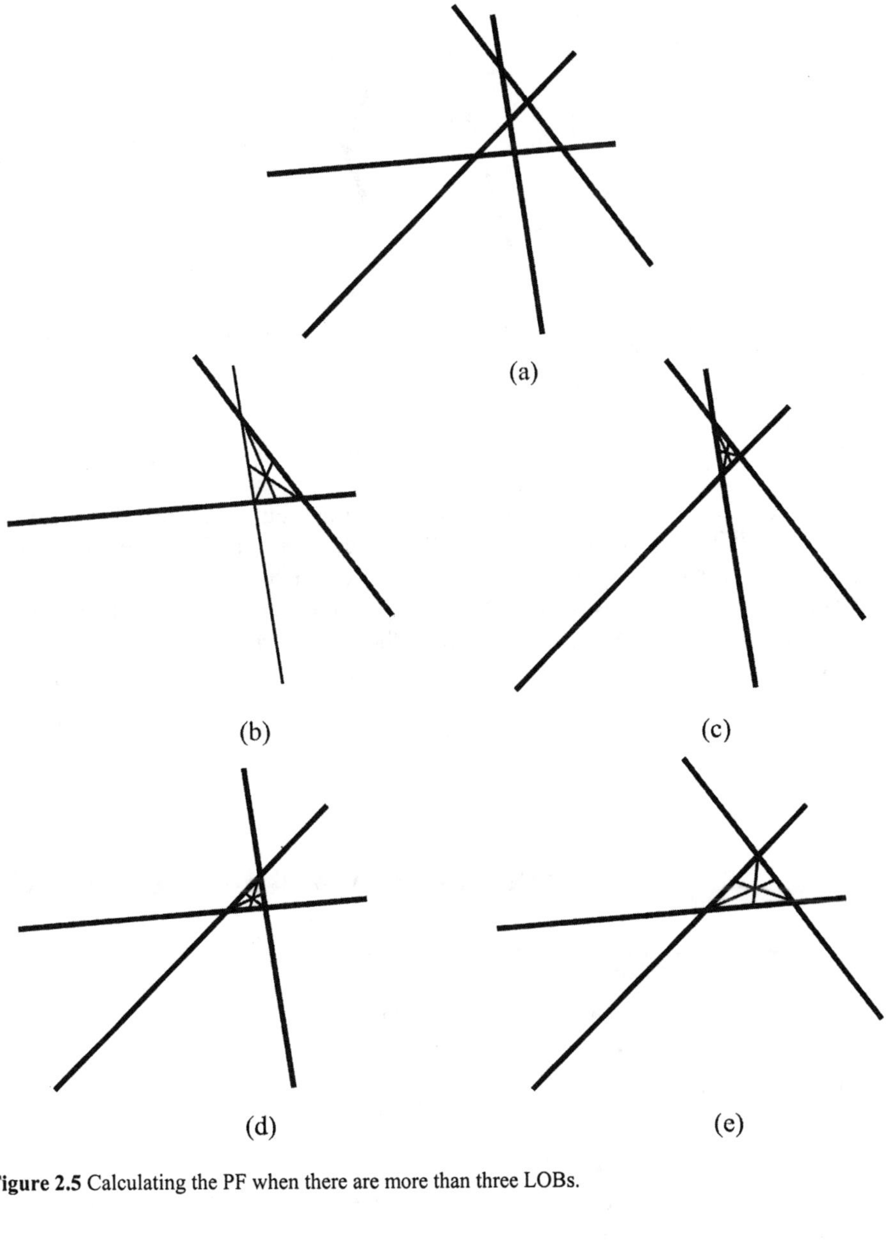

Figure 2.5 Calculating the PF when there are more than three LOBs.

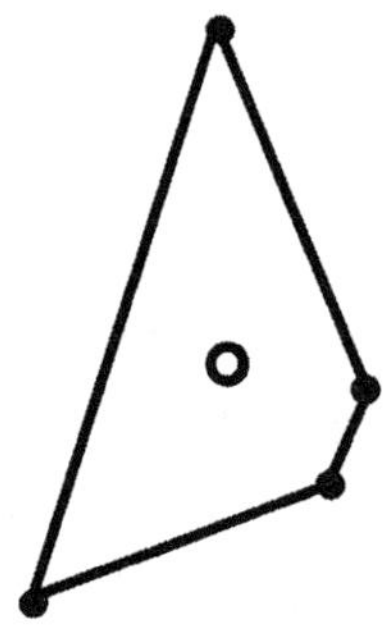

Figure 2.6 The centroid of the polygon formed by connecting the centroids of the triangles formed with three LOBs yields the final fix.

$$c(\hat{\boldsymbol{\theta}}_k) = \tilde{\mathbf{x}}_k^{\mathrm{T}} \mathbf{W}_k \tilde{\mathbf{x}}_k \tag{2.6}$$

where $\mathbf{W}_k$ is a weighting matrix, which is arbitrary but must be symmetric and positive definite. When $\mathbf{W}_k = \mathbf{I}$, then the process is called LSE estimation. If $\mathbf{W}_k \neq \mathbf{I}$ then it is called *weighted LSE* (WLSE) estimation. $\mathbf{W}_k$ may be a constant matrix or it may be updated for each k. The difference between the actual value of $\mathbf{x}_k$ and its estimate $\hat{\mathbf{x}}_k$ at instant k is called the *error term*, denoted by $\tilde{\mathbf{x}}_k$, and is given by

$$\tilde{\mathbf{x}}_k = \mathbf{x}_k - \hat{\mathbf{x}}_k \tag{2.7}$$

and

$$\tilde{\mathbf{x}}_k = [\tilde{x}_k \quad \tilde{x}_{k-1} \quad \cdots \quad \tilde{x}_{k-M+1}]^{\mathrm{T}} \tag{2.8}$$

Substituting (2.7) into (2.6) yields

$$c(\hat{\boldsymbol{\theta}}_k) = (\mathbf{x}_k - \hat{\mathbf{x}}_k)^{\mathrm{T}} \mathbf{W}_k (\mathbf{x}_k - \hat{\mathbf{x}}_k) \tag{2.9}$$

Using (2.5) in (2.9),

$$\begin{aligned} c(\hat{\boldsymbol{\theta}}_k) &= (\mathbf{x}_k - \mathbf{H}_k \hat{\boldsymbol{\theta}}_k)^{\mathrm{T}} \mathbf{W}_k (\mathbf{x}_k - \mathbf{H}_k \hat{\boldsymbol{\theta}}_k) \\ &= (\mathbf{x}_k^{\mathrm{T}} - \hat{\boldsymbol{\theta}}_k^{\mathrm{T}} \mathbf{H}_k^{\mathrm{T}}) \mathbf{W}_k (\mathbf{x}_k - \mathbf{H}_k \hat{\boldsymbol{\theta}}_k) \end{aligned} \tag{2.10}$$

$$
\begin{aligned}
&= (\mathbf{x}_k^{\mathrm{T}}\mathbf{W}_k - \hat{\boldsymbol{\theta}}_k^{\mathrm{T}}\mathbf{H}_k^{\mathrm{T}}\mathbf{W}_k)(\mathbf{x}_k - \mathbf{H}_k\hat{\boldsymbol{\theta}}_k) \\
&= \mathbf{x}_k^{\mathrm{T}}\mathbf{W}_k\mathbf{x}_k - 2\mathbf{x}_k^{\mathrm{T}}\mathbf{W}_k\mathbf{H}_k\hat{\boldsymbol{\theta}}_k + \hat{\boldsymbol{\theta}}_k^{\mathrm{T}}\mathbf{H}_k^{\mathrm{T}}\mathbf{W}_k\mathbf{H}_k\hat{\boldsymbol{\theta}}_k
\end{aligned} \tag{2.11}
$$

This last step is possible because $\mathbf{x}_k^{\mathrm{T}}\mathbf{W}_k\mathbf{H}_k\hat{\boldsymbol{\theta}}_k$ is a scalar [as are all three terms in (2.11)] and therefore symmetric.

The minimum of the cost function is found by finding the derivative of (2.11) and setting it to zero. The first term in (2.11) is independent of $\hat{\boldsymbol{\theta}}_k$, so its derivative is zero. The derivative of the second term is based on a result from vector calculus that says for two $n \times 1$ vectors **a** and **b**,

$$
\frac{d\mathbf{a}^{\mathrm{T}}\mathbf{b}}{d\mathbf{b}} = \mathbf{a} \tag{2.12}
$$

Thus,

$$
\frac{d(-2\mathbf{x}_k^{\mathrm{T}}\mathbf{W}_k\mathbf{H}_k\hat{\boldsymbol{\theta}}_k)}{d\hat{\boldsymbol{\theta}}_k} = -2\left(\mathbf{x}_k^{\mathrm{T}}\mathbf{W}_k\mathbf{H}_k\right)^{\mathrm{T}} = -2\mathbf{H}_k^{\mathrm{T}}\mathbf{W}_k\mathbf{x}_k \tag{2.13}
$$

The derivative of the third term in (2.11) is found by using the fact that for $n \times n$ matrix **M**

$$
\frac{d\mathbf{b}^{\mathrm{T}}\mathbf{M}\mathbf{b}}{d\mathbf{b}} = 2\mathbf{M}\mathbf{b} \tag{2.14}
$$

Therefore,

$$
\frac{d\hat{\boldsymbol{\theta}}_k^{\mathrm{T}}\mathbf{H}_k^{\mathrm{T}}\mathbf{W}_k\mathbf{H}_k\hat{\boldsymbol{\theta}}_k}{d\hat{\boldsymbol{\theta}}_k} = 2\mathbf{H}_k^{\mathrm{T}}\mathbf{W}_k\mathbf{H}_k\hat{\boldsymbol{\theta}}_k \tag{2.15}
$$

Using (2.13) and (2.15),

$$
\frac{dc(\hat{\boldsymbol{\theta}}_k)}{d\hat{\boldsymbol{\theta}}_k} = 0 = -2\mathbf{H}_k^{\mathrm{T}}\mathbf{W}_k\mathbf{x}_k + 2\mathbf{H}_k^{\mathrm{T}}\mathbf{W}_k\mathbf{H}_k\hat{\boldsymbol{\theta}}_k \tag{2.16}
$$

so

$$\hat{\boldsymbol{\theta}}_k = [\mathbf{H}_k^{\mathrm{T}} \mathbf{W}_k \mathbf{H}_k]^{-1} \mathbf{H}_k^{\mathrm{T}} \mathbf{W}_k \mathbf{x}_k \tag{2.17}$$

which is the optimal LSE estimate desired. Expression (2.17) is sometimes referred to as the *normal equation.* That this is a minimum can be established by taking the second derivative of (2.11), which is

$$\frac{d^2 c(\hat{\boldsymbol{\theta}}_k)}{d\hat{\boldsymbol{\theta}}_k^2} = 2\mathbf{H}_k^{\mathrm{T}} \mathbf{W}_k \mathbf{H}_k \tag{2.18}$$

which is in quadratic form and, since $\mathbf{W}_k$ is positive definite, $\mathbf{H}_k^{\mathrm{T}} \mathbf{W}_k \mathbf{H}_k > 0$, and $\hat{\boldsymbol{\theta}}_k$ is a minimum.

The accuracy or quality of the estimate provided by (2.17) is given by the covariance matrix, which is

$$\mathbf{Q}_k = [\mathbf{H}_k^T \mathbf{W}_k^{-1} \mathbf{H}_k]^{-1} \tag{2.19}$$

In three dimensions, when the covariances are fixed, this matrix is (k assumed)

$$\mathbf{Q} = \begin{bmatrix} \sigma_x^2 & \rho_{xy}\sigma_x\sigma_y & \rho_{xz}\sigma_x\sigma_z \\ \rho_{xy}\sigma_x\sigma_y & \sigma_y^2 & \rho_{yz}\sigma_y\sigma_z \\ \rho_{xz}\sigma_x\sigma_z & \rho_{yz}\sigma_y\sigma_z & \sigma_z^2 \end{bmatrix} \tag{2.20}$$

where σ_x^2 is the variance of x, ρ_{xy} is the correlation coefficient between x and y, and so forth. It is often assumed that x, y, and z are uncorrelated so that $\rho_{xy} = \rho_{xz} = \rho_{yz} = 0$, in which case

$$\mathbf{Q} = \begin{bmatrix} \sigma_x^2 & 0 & 0 \\ 0 & \sigma_y^2 & 0 \\ 0 & 0 & \sigma_z^2 \end{bmatrix} \tag{2.21}$$

The nonrecursive form of the LSE estimation procedure is identical to the above except that k is not the sample time but is interpreted as the size of the final measurement set.

Example: Suppose it is desired to fit a set of N paired points (x_i, y_i) to a straight line. The equation for the line is

$$y = \beta_0 + \beta_1 x \tag{2.22}$$

The problem can be cast into matrix form as

$$\mathbf{y} = \begin{bmatrix} y_1 & y_2 & \cdots & y_N \end{bmatrix}^{\mathrm{T}} \tag{2.23}$$

$$\boldsymbol{\beta} = \begin{bmatrix} \beta_0 & \beta_1 \end{bmatrix}^{\mathrm{T}} \tag{2.24}$$

$$\mathbf{x} = \begin{bmatrix} 1 & 1 & \cdots & 1 \\ x_1 & x_2 & \cdots & x_N \end{bmatrix}^{\mathrm{T}} \tag{2.25}$$

The LSE estimate based on (2.17) is given by

$$\hat{\boldsymbol{\theta}} = [\mathbf{x}^{\mathrm{T}} \mathbf{W}\, \mathbf{x}\,]^{-1} \mathbf{x}^{\mathrm{T}} \mathbf{W}\, \mathbf{y} \tag{2.26}$$

where **W** = **I**, yielding

$$\hat{\boldsymbol{\theta}} = [\mathbf{x}^{\mathrm{T}} \mathbf{x}\,]^{-1} \mathbf{x}^{\mathrm{T}} \mathbf{y} \tag{2.27}$$

Now,

$$\mathbf{x}^{\mathrm{T}} \mathbf{x} = \begin{bmatrix} 1 & 1 & \cdots & 1 \\ x_1 & x_2 & \cdots & x_N \end{bmatrix} \begin{bmatrix} 1 & x_1 \\ 1 & x_2 \\ \vdots & \vdots \\ 1 & x_N \end{bmatrix} \tag{2.28}$$

$$= \begin{bmatrix} N & \sum_{i=1}^{N} x_i \\ \sum_{i=1}^{N} x_i & \sum_{i=1}^{N} x_i^2 \end{bmatrix} \tag{2.29}$$

so

$$\left[\mathbf{x}^{\mathrm{T}} \mathbf{x}\right]^{-1} = \frac{1}{N \sum_{i=1}^{N} x_i^2 - \left(\sum_{i=1}^{N} x_i\right)^2} \begin{bmatrix} \sum_{i=1}^{N} x_i^2 & -\sum_{i=1}^{N} x_i \\ -\sum_{i=1}^{N} x_i & N \end{bmatrix} \tag{2.30}$$

and

$$\mathbf{x}^{\mathrm{T}}\mathbf{y} = \begin{bmatrix} \sum_{i=1}^{N} y_i \\ \sum_{i=1}^{N} x_i y_i \end{bmatrix} \tag{2.31}$$

From (2.27),

$$\begin{aligned}\hat{\boldsymbol{\beta}} = \begin{bmatrix} \hat{\beta}_0 \\ \hat{\beta}_1 \end{bmatrix} &= \frac{1}{N\sum_{i=1}^{N} x_i^2 - \left(\sum_{i=1}^{N} x_i\right)^2} \begin{bmatrix} \sum_{i=1}^{N} x_i^2 & -\sum_{i=1}^{N} x_i \\ -\sum_{i=1}^{N} x_i & N \end{bmatrix} \begin{bmatrix} \sum_{i=1}^{N} y_i \\ \sum_{i=1}^{N} x_i y_i \end{bmatrix} \\ &= \frac{1}{N\sum_{i=1}^{N} x_i^2 - \left(\sum_{i=1}^{N} x_i\right)^2} \begin{bmatrix} \sum_{i=1}^{N} y_i \sum_{i=1}^{N} x_i^2 - \sum_{i=1}^{N} x_i \sum_{i=1}^{N} x_i y_i \\ N\sum_{i=1}^{N} x_i y_i - \sum_{i=1}^{N} x_i \sum_{i=1}^{N} y_i \end{bmatrix}\end{aligned} \tag{2.32}$$

which are the optimal estimates of the coefficients in an LSE sense.

■

Substituting (2.3) into (2.17) with $\mathbf{W} = \mathbf{I}$ (and dropping the k) yields

$$\hat{\boldsymbol{\theta}} = (\mathbf{H}^{\mathrm{T}}\mathbf{H})^{-1}\mathbf{H}^{\mathrm{T}}(\mathbf{H}\boldsymbol{\theta} + \mathbf{n}) \tag{2.33}$$

$$= \underbrace{(\mathbf{H}^{\mathrm{T}}\mathbf{H})^{-1}\mathbf{H}^{\mathrm{T}}\mathbf{H}}_{=\mathbf{I}}\boldsymbol{\theta} + (\mathbf{H}^{\mathrm{T}}\mathbf{H})^{-1}\mathbf{H}^{\mathrm{T}}\mathbf{n} \tag{2.34}$$

so

$$\hat{\boldsymbol{\theta}} = \boldsymbol{\theta} + (\mathbf{H}^{\mathrm{T}}\mathbf{H})^{-1}\mathbf{H}^{\mathrm{T}}\mathbf{n} \tag{2.35}$$

The *bias error* is given by

$$\boldsymbol{\delta} = \mathcal{E}\{\hat{\boldsymbol{\theta}}\} - \boldsymbol{\theta} \tag{2.36}$$

The resulting LSE estimation bias is approximated by [3]

$$\boldsymbol{\delta} \approx \mathcal{E}\left\{\frac{1}{N}\mathbf{H}^{\mathrm{T}}\mathbf{H}\right\}\mathcal{E}\left\{\mathbf{H}^{\mathrm{T}}\mathbf{n}\right\} \tag{2.37}$$

Therefore, in general the LSE estimation process produces biased estimates. However, specific cases of LSE estimation, as will be shown, can produce unbiased results.

2.3 Total Least-Squared Estimation

A generalization of the LSE estimation technique discussed in Section 2.2 allows for the existence of noise in not only the measurements $\mathbf{x}_k$ in (2.3), but also for errors in the observation matrix $\mathbf{H}_k$. When there is no noise in $\mathbf{H}_k$, and the noise represented by $\mathbf{n}_k \sim \mathcal{N}(0,\sigma^2)$ (this notation means that the noise is normal, or Gaussian, with zero mean and variance σ^2), then the LSE solution $\hat{\boldsymbol{\theta}}_{k,\mathrm{LS}}$ is the same as the maximum likelihood estimate. However, when noise is present in $\mathbf{H}_k$, (2.17) with $\mathbf{W}_k = \mathbf{I}$ is no longer optimal. It exhibits bias and increased covariance. To determine the optimal LSE estimate in this case, the method of TLSE estimation was developed [4]. The subscript k will be dropped here for notational convenience.

As described in Section 2.2, the LSE estimate is obtained by finding $\hat{\boldsymbol{\theta}}$ by calculating

$$\min_{\boldsymbol{\theta}} \|\mathbf{H}\boldsymbol{\theta} - \mathbf{x}\|_2 \tag{2.38}$$

where $\|\mathbf{z}\|_2$ is the L_2 norm of $\mathbf{z}$. The solution to (2.38) is given by

$$\hat{\boldsymbol{\theta}}_{\mathrm{LS}} = \mathbf{H}^{\dagger}\mathbf{x} \tag{2.39}$$

where $\mathbf{H}^{\dagger}$ is the pseudoinverse of $\mathbf{H}_{M\times P}$.[1] It is typically assumed that $M > P$ and that $\mathbf{H}$ has full rank, so that $\mathbf{H}^{\dagger} = (\mathbf{H}^*\mathbf{H})^{-1}\mathbf{H}^*$ and

$$\hat{\boldsymbol{\theta}}_{\mathrm{LS}} = (\mathbf{H}^{\mathrm{T}}\mathbf{H})^{-1}\mathbf{H}^*\mathbf{x} \tag{2.40}$$

The measurement vector and observation matrix can be expressed as

---

[1] For any full rank matrix $\mathbf{A}_{(m\times n)}$ $(m > n)$ (rank $= n$) the *pseudoinverse* of $\mathbf{A}$ is $\mathbf{A}^{\dagger} = (\mathbf{A}^{\mathrm{T}}\mathbf{A})^{-1}\mathbf{A}^{\mathrm{T}}$. One way to compute $\mathbf{A}^{\dagger}$ is to use *singular value decomposition* (SVD). With $\mathbf{A} = \mathbf{U}\mathbf{S}\mathbf{V}^{\mathrm{T}}$, where $\mathbf{U}_{(m\times m)}$ and $\mathbf{V}_{(n\times n)}$ are orthogonal and $\mathbf{S}_{(m\times n)} = \begin{bmatrix}\mathbf{D}_{(n\times n)} & \mathbf{0}_{(m-n\times n)}\end{bmatrix}^{\mathrm{T}}$, with $\mathbf{D}_{(n\times n)}$ diagonal with real, nonnegative singular values d_{ii}, then $\mathbf{A}^{\dagger} = \mathbf{V}(\mathbf{S}^{\mathrm{T}}\mathbf{S})^{-1}\mathbf{S}^{\mathrm{T}}\mathbf{U}^{\mathrm{T}}$. If $r = \mathrm{rank}(\mathbf{A}) < n$, the inverse of $\mathbf{S}^{\mathrm{T}}\mathbf{S}$ does not exist, and the first r singular values only are used in $\mathbf{S}$, and $\mathbf{U}$ and $\mathbf{V}$ are correspondingly smaller.

$$\mathbf{x} = \mathbf{x}_0 + \Delta\mathbf{x} \qquad \mathbf{H} = \mathbf{H}_0 + \Delta\mathbf{H} \tag{2.41}$$

where $\Delta\mathbf{x}$ and $\Delta\mathbf{H}$ are the noise perturbations. With no noise, $\mathbf{H}_0\boldsymbol{\theta}_0 = \mathbf{x}_0$. The TLSE estimate is obtained from

$$\min_{\Delta\mathbf{D},\boldsymbol{\theta}} \|\Delta\mathbf{D}\|_\mathrm{F}^2 \qquad \text{subject to} \qquad (\mathbf{D} + \Delta\mathbf{D})\begin{pmatrix} \boldsymbol{\theta} \\ -1 \end{pmatrix} = \mathbf{0} \tag{2.42}$$

$\mathbf{D}$ is given by

$$\mathbf{D} = [\mathbf{H} : \mathbf{x}] \tag{2.43}$$

and $\Delta\mathbf{D}$ is given by

$$\Delta\mathbf{D} = [\Delta\mathbf{H} : \Delta\mathbf{x}] \tag{2.44}$$

The value $\|\mathbf{z}\|_\mathrm{F}$ stands for the *Frobenius norm* of $\mathbf{z}$, which is given by $\|\mathbf{z}\|_\mathrm{F} = \sqrt{\mathrm{tr}\left(\mathbf{z}^*\mathbf{z}\right)}$. Let the SVD of $\mathbf{D}$ be $\mathbf{D} = \mathbf{U\Sigma V}^*$ ($M \times P$), where $\boldsymbol{\Sigma}$ has real diagonal elements σ_i such that

$$\sigma_1 \geq \sigma_2 \geq \ldots \geq \sigma_{\min(M,P)} \geq 0, \qquad i = 1, 2, \ldots, M+1 \tag{2.45}$$

The σ_i are the singular values of $\mathbf{D}$ and the first min(M, P) columns of $\mathbf{U}$ and $\mathbf{V}$ are the left and right singular vectors of $\mathbf{D}$. These singular values and singular vectors satisfy

$$\mathbf{D}\mathbf{v}_i = \sigma_i \mathbf{u}_i \qquad \text{and} \qquad \mathbf{D}^*\mathbf{u}_i = \sigma_i \mathbf{v}_i \tag{2.46}$$

where $\mathbf{u}_i$ and $\mathbf{v}_i$ are the ith columns of $\mathbf{U}$ and $\mathbf{V}$, respectively.

Based on these derivations, the TLS estimate is given by

$$\hat{\boldsymbol{\theta}} = -\frac{1}{v_l}\mathbf{v}' \tag{2.47}$$

or

$$\hat{\boldsymbol{\theta}}_\mathrm{TLS} = (\mathbf{D}^*\mathbf{D} - \sigma_i^2\mathbf{I})^{-1}\mathbf{D}^*\mathbf{x} \tag{2.48}$$

The factor v_l is the last component of $\mathbf{V}$ in the ith row and $\mathbf{v}'$ is the first element of $\mathbf{V}$. In (2.48), σ_i is the smallest singular value and $\sigma_i^2\mathbf{I}$ helps to reduce the bias that the noise in the observation matrix introduces in $\mathbf{D}^*\mathbf{D}$.

When $\Delta\mathbf{D} \sim \mathcal{N}(0,\sigma_{\mathbf{D}}^2)$, TLSE estimation is very effective and more accurate than LSE.

2.4 Least-Squared Distance Error PF Algorithm

2.4.1 Brown's Least-Squared Triangulation Algorithm

An algorithm developed by Brown [5] will be presented in this section for calculating the PF, which is based on minimizing the square of the miss distance of the PF from the measured LOBs. This algorithm was presented in [6] and is included here for completeness

Referring to Figure 2.7, to minimize the sum of the squares of the total miss distance, formulate

$$D = \sum_{i=1}^{N} d_i^2 \tag{2.49}$$

$$\begin{aligned} &= \sum_{i=1}^{N} a_i^2 x_\mathrm{T}^2 + \sum_{i=1}^{N} 2a_i b_i x_\mathrm{T} y_\mathrm{T} - \sum_{i=1}^{N} 2a_i c_i x_\mathrm{T} \\ &\quad + \sum_{i=1}^{N} b_i^2 y_\mathrm{T}^2 - \sum_{i=1}^{N} 2b_i c_i y_\mathrm{T} + \sum_{i=1}^{N} c_i^2 \end{aligned} \tag{2.50}$$

where

$$\begin{aligned} a_i &= \sin\phi_i \\ b_i &= -\cos\phi_i \\ c_i &= x_i \sin\phi_i - y_i \cos\phi_i \\ N &= \text{number of LOBs} \end{aligned}$$

Setting the first partial derivative of D with respect to x_T and then y_T equal to zero will find the values of x_T and y_T for which the total squared distance is minimized.

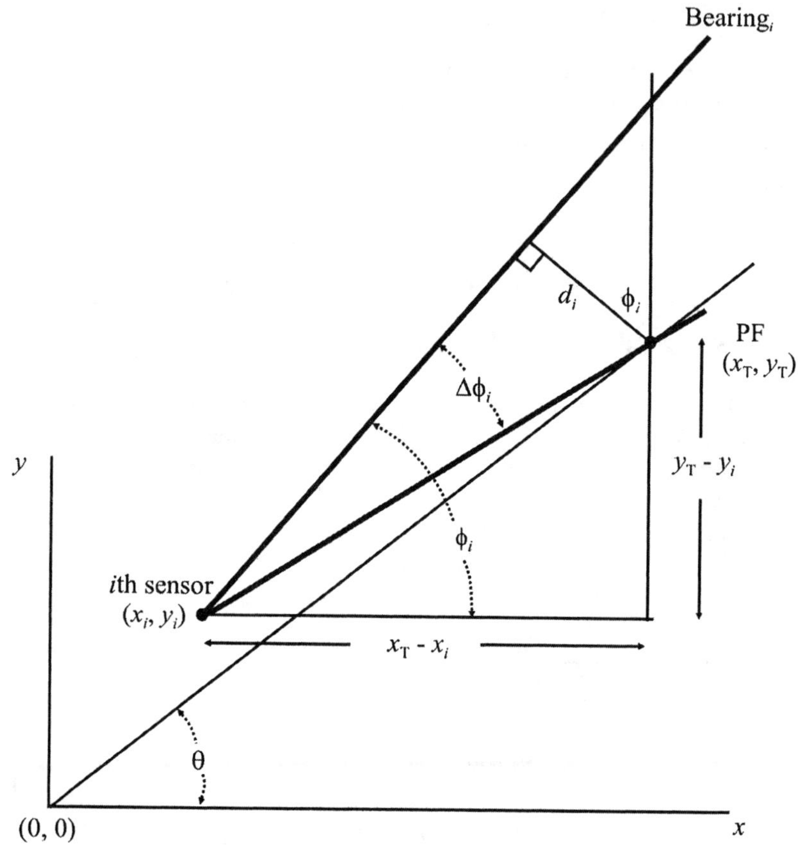

Figure 2.7 Definitions of the terms for derivation of Brown's mean-squared distance algorithm. (*Source:* [5].)

$$\frac{\partial D}{\partial x_T} = 0 = 2x_T \sum_{i=1}^{N} a_i^2 + 2y_T \sum_{i=1}^{N} a_i b_i - 2\sum_{i=1}^{N} a_i c_i \tag{2.51}$$

$$\frac{\partial D}{\partial y_T} = 0 = 2x_T \sum_{i=1}^{N} a_i b_i + 2y_T \sum_{i=1}^{N} b_i^2 - 2\sum_{i=1}^{N} b_i c_i \tag{2.52}$$

which yield

$$x_{\mathrm{T}} = \frac{\sum_{i=1}^{N} b_i^2 \sum_{i=1}^{N} a_i c_i - \sum_{i=1}^{N} a_i b_i \sum_{i=1}^{N} b_i c_i}{\sum_{i=1}^{N} a_i^2 \sum_{i=1}^{N} b_i^2 - \left(\sum_{i=1}^{N} a_i b_i \right)^2} \tag{2.53}$$

$$y_{\mathrm{T}} = \frac{\sum_{i=1}^{N} a_i^2 \sum_{i=1}^{N} b_i c_i - \sum_{i=1}^{N} a_i b_i \sum_{i=1}^{N} a_i c_i}{\sum_{i=1}^{N} a_i^2 \sum_{i=1}^{N} b_i^2 - \left(\sum_{i=1}^{N} a_i b_i \right)^2} \tag{2.54}$$

The above miss distance for sensor i is expressed as

$$d_i = a_i x_{\mathrm{T}} + b_i y_{\mathrm{T}} - c_i \tag{2.55}$$

where i is the ith measurement of a line of bearing, and a_i, b_i, and c_i are as given above. In matrix form this is

$$\mathbf{D} = \mathbf{H}\mathbf{x}_{\mathrm{T}} - \mathbf{C} \tag{2.56}$$

In this expression,

$$\mathbf{C} = \begin{bmatrix} c_1 \\ c_2 \\ \vdots \\ c_N \end{bmatrix} \qquad \mathbf{x}_{\mathrm{T}} = \begin{bmatrix} x_{\mathrm{T}} \\ y_{\mathrm{T}} \end{bmatrix} \qquad \mathbf{H} = \begin{bmatrix} a_1 & b_1 \\ a_2 & b_2 \\ \vdots & \vdots \\ a_N & b_N \end{bmatrix} \qquad \mathbf{D} = \begin{bmatrix} d_1 \\ d_2 \\ \vdots \\ d_N \end{bmatrix} \tag{2.57}$$

The LSE estimator for the target location vector $\mathbf{x}_{\mathrm{T}}$ is given by (2.17) as

$$\mathbf{x}_{\mathrm{T}} = \left[\mathbf{H}^{\mathrm{T}} \mathbf{W}^{-1} \mathbf{H} \right]^{-1} \mathbf{H}^{\mathrm{T}} \mathbf{W}^{-1} \mathbf{C} \tag{2.58}$$

where, as usual, the superscript $^{-1}$ denotes inverse and ^{T} denotes transpose. We assume that the noise is zero mean AWGN. The variance of this estimator is given by the two-dimensional covariance matrix (2.19)

$$\mathbf{Q} = \left[\mathbf{H}^{\mathrm{T}}\mathbf{W}^{-1}\mathbf{H}\right]^{-1} = \begin{bmatrix} \sigma_x^2 & \rho_{xy}\sigma_x\sigma_y \\ \rho_{xy}\sigma_x\sigma_y & \sigma_y^2 \end{bmatrix} \tag{2.59}$$

The EEP parameters are related to the elements of this covariance matrix as follows:

$$\begin{aligned} L_{\mathrm{A}} &= \text{semimajor axis} \\ &= \frac{2\left(\sigma_x^2\sigma_y^2 - \rho_{xy}^2\sigma_x^2\sigma_y^2\right)C^2}{\sigma_x^2 + \sigma_y^2 - \left[\left(\sigma_y^2 - \sigma_x^2\right)^2 + 4\rho_{xy}^2\sigma_x^2\sigma_y^2\right]^{1/2}} \end{aligned} \tag{2.60}$$

$$\begin{aligned} L_{\mathrm{I}} &= \text{semiminor axis} \\ &= \frac{2\left(\sigma_x^2\sigma_y^2 - \rho_{xy}^2\sigma_x^2\sigma_y^2\right)C^2}{\sigma_x^2 + \sigma_y^2 + \left[\left(\sigma_y^2 - \sigma_x^2\right)^2 + 4\rho_{xy}^2\sigma_x^2\sigma_y^2\right]^{1/2}} \end{aligned} \tag{2.61}$$

$$\tan\ 2\theta = \frac{2\rho_{xy}^2\sigma_x\sigma_y}{\sigma_y^2 - \sigma_x^2} \tag{2.62}$$

$$C = -2\ \ln(1 - P_e) \tag{2.63}$$

$$P_e = \text{probability of being inside} \tag{2.64}$$

Here, θ is the tilt angle of the semimajor axis of the ellipse relative to the x-axis.

The weighting matrix $\mathbf{W}^{-1}$ is used to optimize the performance. In one application of this algorithm, $\mathbf{W}^{-1}$ is given by

$$\mathbf{W}^{-1} = \frac{1}{\sum_i QF_i} \begin{bmatrix} QF_1 & 0 & 0 & \cdots & 0 \\ 0 & QF_2 & 0 & \cdots & 0 \\ 0 & 0 & QF_3 & \cdots & 0 \\ \vdots & \vdots & \vdots & \ddots & \vdots \\ 0 & 0 & 0 & \cdots & QF_N \end{bmatrix} \times \begin{bmatrix} \frac{1}{\sigma_{d_1}^2} & 0 & 0 & \cdots & 0 \\ 0 & \frac{1}{\sigma_{d_2}^2} & 0 & \cdots & 0 \\ 0 & 0 & \frac{1}{\sigma_{d_3}^2} & \cdots & 0 \\ \vdots & \vdots & \vdots & \ddots & \vdots \\ 0 & 0 & 0 & \cdots & \frac{1}{\sigma_{d_N}^2} \end{bmatrix} \tag{2.65}$$

where QF_i is some quality factor associated with measurement i. It might be the variance of the measurement, it could be some higher order statistic as it makes sense, or it could be a measure of the *signal-to-noise ratio* (SNR), as examples.

2.4.1.1 Mean and Covariance of the Bias in Brown's Algorithm

Matrix **H** is given by (2.57) which contains no random components. Therefore, $\mathbf{H}^{\mathrm{T}}\mathbf{H}$ contains no random components. Thus, from (2.37),

$$\boldsymbol{\delta} \approx -\frac{1}{N^2}(\mathbf{H}^{\mathrm{T}}\mathbf{H})^{-1}\mathbf{H}^{\mathrm{T}}\mathcal{E}\{\mathbf{n}\} \tag{2.66}$$

Since it is assumed that $\mathbf{n} \sim \mathcal{N}(\mathbf{0},\sigma^2)$, $\mathcal{E}\{\mathbf{n}\} = 0$ and therefore $\boldsymbol{\delta} = 0$.

Denoting the covariance matrix of **D** by **V** and assuming for simplicity that $\mathbf{W} = \mathbf{I}$, the variance of the estimator described by (2.58) is given by [7]

$$\mathbf{Cov}(\mathbf{C}) = (\mathbf{H}^{\mathrm{T}}\mathbf{H})^{-1}\mathbf{H}^{\mathrm{T}}\mathbf{V}\mathbf{H}(\mathbf{H}^{\mathrm{T}}\mathbf{H})^{-1} \tag{2.67}$$

2.4.2 Hemispheric Least-Squared Error Estimation Algorithm

A different least-squared algorithm for PF estimation is presented in this section. The measured bearing from a sensor, ϕ_i, as illustrated in Figure 2.8, projected onto the surface of the Earth in the northern hemisphere (the other hemisphere will work as well with appropriate change of variables) can be characterized by [8,9]

$$\cos(\phi_i) = \frac{\psi_{\mathrm{T}} - \psi_i}{\sqrt{(\psi_{\mathrm{T}} - \psi_i)^2 + (\lambda_{\mathrm{T}} - \lambda_i)^2}} \tag{2.68}$$

where the variables are defined in Figure 2.8 as (all in degrees)

λ_i = longitude of the sensor during the ith observation interval
ψ_i = latitude of the sensor during the ith observation interval
λ_{T} = actual longitude of the target emitter
ψ_{T} = actual latitude of the target emitter
η = heading of the sensor relative to north
ϕ_i = ith measured bearing
$\Delta\phi_i$ = error in the ith measured bearing
$\Delta\lambda_i$ = error in the calculation of the longitude in the ith observation interval
$\Delta\psi_i$ = error in the calculation of the latitude in the ith observation interval

It is assumed that all LOBs are referenced to north; therefore, adjustments for vehicle/aircraft navigation/orientation/attitude have been previously accounted for. It is further assumed that N LOB measurements are made at regular intervals as indicated by the index i.

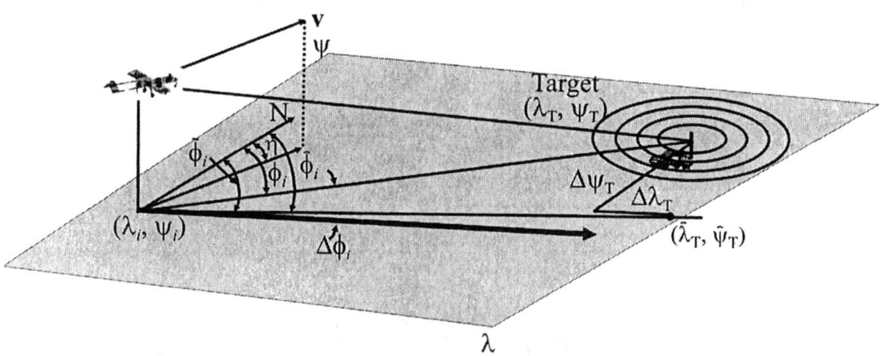

Figure 2.8 Geometry for hemispheric least-squared.

If the bearing errors are normally distributed with zero mean, the best least-squared estimates of the PF, $(\hat{\lambda}_T, \hat{\psi}_T)$, assuming that ($\lambda_0$, ψ_0) is available or estimated, are given by

$$\lambda_i = \hat{\lambda}_i - \Delta\lambda_i \tag{2.69}$$

and

$$\psi_i = \hat{\psi}_i - \Delta\psi_i \tag{2.70}$$

Using (1.7), expanding ($\lambda_i - \Delta\lambda_i$) about $\Delta\lambda_T = \Delta\psi_T = 0$ yields

$$Y_i = X_{i1}\Delta\lambda_T + X_{i2}\Delta\psi_T + \epsilon_i \tag{2.71}$$

where

$$\begin{aligned}
Y_i &= \tilde{\phi}_i - \hat{\phi}_i \\
X_{i1} &= \frac{\hat{n}}{\hat{m}^2 + \hat{n}^2} \\
X_{i2} &= \frac{\hat{m}}{\hat{m}^2 + \hat{n}^2} \\
\hat{\phi}_i &= \cos^{-1}\left(\frac{\hat{m}}{\sqrt{\hat{m}^2 + \hat{n}^2}}\right) \\
\hat{m} &= \hat{\lambda}_T - \lambda_i \\
\hat{n} &= (\hat{\psi}_T - \psi_i) \\
\epsilon_i &= \Delta\phi_i + \text{higher order terms}
\end{aligned}$$

If there is systematic error manifest in a bias in the measurements given by ϕ_s, then

$$\Delta\phi_i = \phi_s + \Delta\phi_i^* \tag{2.72}$$

where $\Delta\phi_i^*$ is the random part of $\Delta\phi_I$, and (2.71) is not necessarily the best estimator.

This procedure yields an estimate of the location of the target according to the expression

$$\mathbf{Y} = \mathbf{X}\boldsymbol{\beta} + \mathbf{E} \tag{2.73}$$

where

$$\mathbf{X} = \begin{bmatrix} X_{11} & X_{21} & \cdots & X_{N1} \\ X_{12} & X_{22} & \cdots & X_{N2} \end{bmatrix}^{\mathrm{T}} \tag{2.74}$$

$$\boldsymbol{\beta} = \begin{bmatrix} \Delta\phi_{\mathrm{T}} & \Delta\lambda_{\mathrm{T}} \end{bmatrix}^{\mathrm{T}} \tag{2.75}$$

$$\mathbf{Y} = \tilde{\boldsymbol{\varphi}} - \hat{\boldsymbol{\varphi}} \tag{2.76}$$

and

$$\mathbf{E} = \begin{bmatrix} E_1 & E_2 & \cdots & E_N \end{bmatrix}^{\mathrm{T}} \tag{2.77}$$

The estimate of $\boldsymbol{\beta}$ is obtained by solving (2.73) for $\boldsymbol{\beta}$ as in (2.17) with $\mathbf{W} = \mathbf{I}$:

$$\boldsymbol{\beta} = \left(\mathbf{X}^{\mathrm{T}}\mathbf{X}\right)^{-1}\mathbf{X}^{\mathrm{T}}\mathbf{Y} = \begin{bmatrix} \hat{\beta}_1 & \hat{\beta}_2 \end{bmatrix}^{\mathrm{T}} \tag{2.78}$$

where $\hat{\beta}_1$ and $\hat{\beta}_2$ are the best estimate of the errors in calculating the PF of the target. The target location is thus obtained by subtracting these values from the initial estimates given by λ_0 and ψ_0 as

$$\hat{\psi}_{\mathrm{T}} = \psi_0 - \beta_1 \tag{2.79}$$

and

$$\hat{\lambda}_{\mathrm{T}} = \lambda_0 - \beta_2 \tag{2.80}$$

This is one iteration through the procedure. It can be repeated to generate better estimates as many times as desired by using the computed values of $\hat{\psi}_{\mathrm{T}}$ and $\hat{\lambda}_{\mathrm{T}}$ from (2.79) and (2.80), respectively, as the initial values of the next iteration, so that

$$\psi_0 \leftarrow \hat{\psi}_{\mathrm{T}} \tag{2.81}$$

and

$$\lambda_0 \leftarrow \hat{\lambda}_T \tag{2.82}$$

For the same reasons discussed above for Brown's algorithm, since the mean of the noise is zero, the resulting bias due to the algorithm is zero as well.

A MathCad® program is given in the appendix to this chapter to compute one pass of this technique.

Example: Consider the case when the target is located at (15, 15) units and the sensor is an aircraft flying along the x-axis as indicated in Figure 2.9. There are 21 measurements taken at unit intervals. The bearings are corrupted by Gaussian noise with a variance of 4 degrees squared.

A few results of estimating the PF of the target with the algorithm described in this section are shown in Figure 2.10. Although the cases illustrated in Figure 2.10 all appear to be converging to the target location, they do so slowly. Convergence was not obtained in all cases. When the initial points were (5, 10), (1, 1), and (10, 20), the algorithm diverged, for example.

For those cases illustrated in Figure 2.10, more than 10 iterations are required to have the results converge. If the initial point is closer to the actual target location, then convergence is faster, although it is still not assured. The results when the initial point was (13, 13) are illustrated in Figure 2.11. While convergence appears to occur, it is still fairly slow.

■

A well-known problem with the least-squared approach to PF is that an initial estimate of the emitter location is required and it is only valid for small bearing errors. Convergence is not guaranteed, as illustrated in the previous example. Also, slow convergence may occur even when convergence occurs. A technique using total least-squared is described in Section 2.4.4 that does not have these shortcomings.

Since the bearing errors were assumed to have zero mean, the covariance matrix of the estimate, assuming no weighting is applied, is given by

$$\mathbf{Cov} = (\mathbf{H}^T\mathbf{H})^{-1}\mathbf{H}^T\mathbf{V}\mathbf{H}(\mathbf{H}^T\mathbf{H})^{-1} \tag{2.83}$$

where $\mathbf{V}$ is the covariance matrix of $\mathbf{E}$.

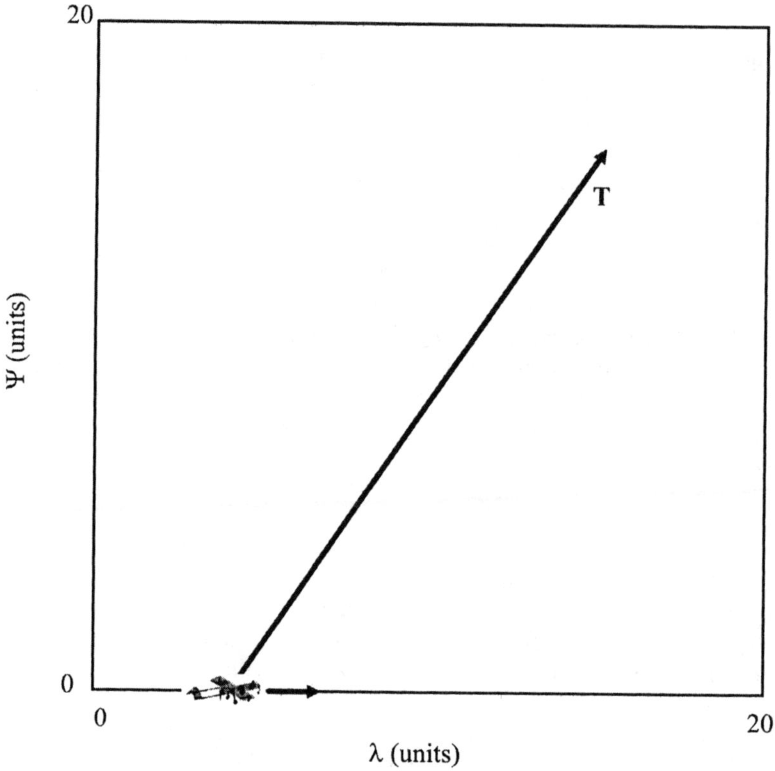

Figure 2.9 Scenario for the least-squared example. The sensor is flying across the x-axis while the target is located at (15, 15).

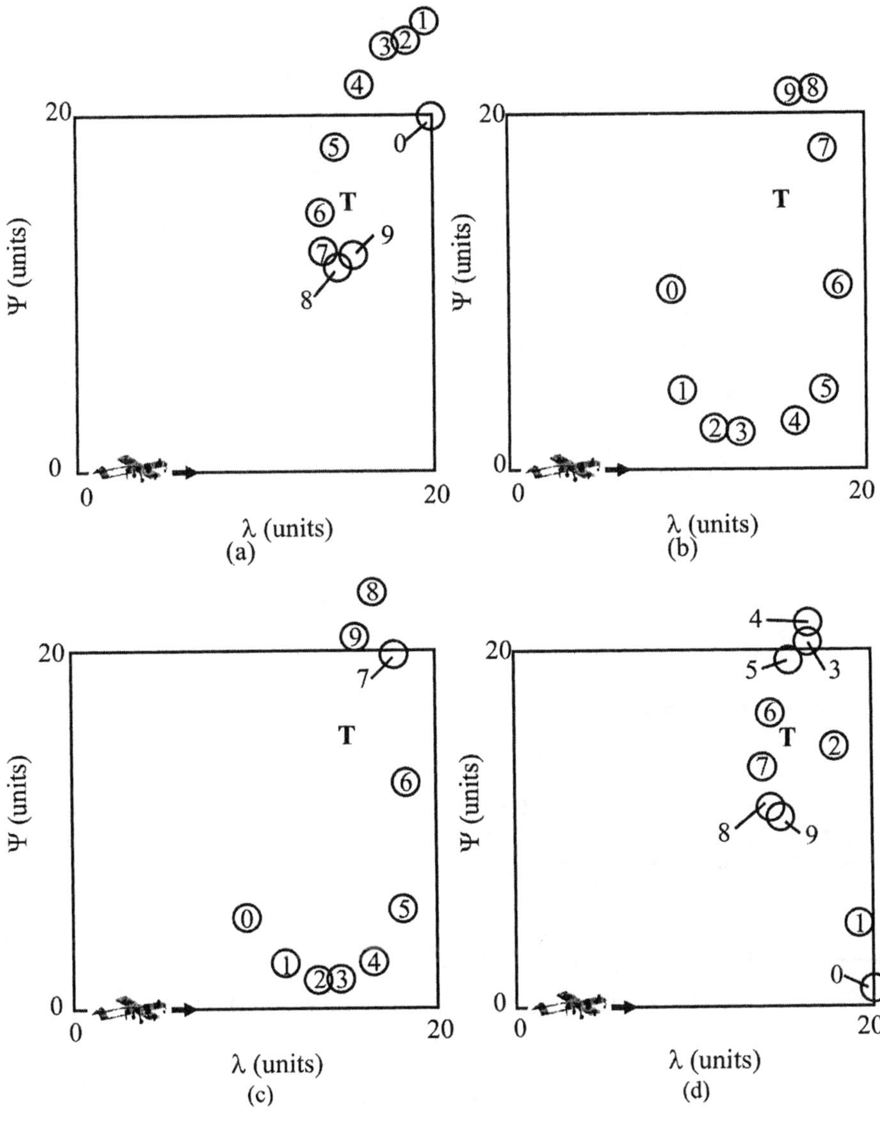

Figure 2.10 Illustrative results of estimating the PF of the target at coordinates (15, 15) with the least-squared algorithm. In (a) the initial point was (20, 20), in (b) it was (10, 10), in (c) it was at (10, 5), and in (d) it was at (20, 1).

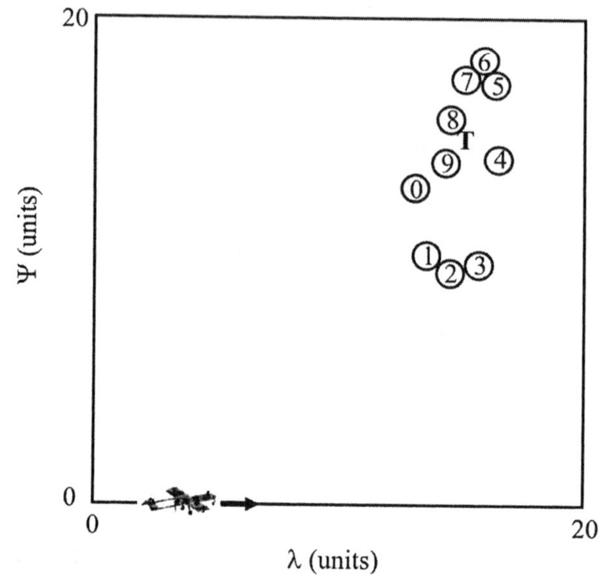

Figure 2.11 Results of LSE when the initial point was closer to the target location at (13, 13).

2.4.3 Pages-Zamora Least-Squared

Pages-Zamora, Vidal, and Brooks [10] presented an LSE algorithm based on the cellular phone requirements imposed by the FCC in the United States. The fundamental algorithm is based on the geometry shown in Figure 2.12 and recognition of the fact that

$$\mathbf{r} = \mathbf{r}_i + d_i \mathbf{v}_i \qquad \forall i \tag{2.84}$$

where d_i is the distance between the target and sensor i and $\mathbf{v}_i$ is the unitary vector given by

$$\mathbf{v}_i = \begin{bmatrix} \cos\phi_i & \sin\phi_i \end{bmatrix}^{\mathrm{T}} \tag{2.85}$$

The distance d_i can be removed from (2.84) as follows:

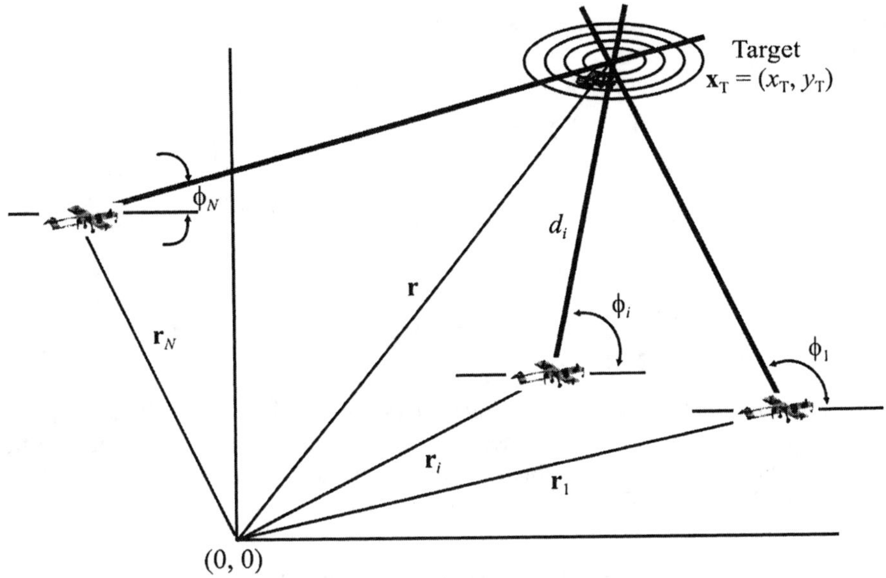

Figure 2.12 Geometry describing the algorithm developed by Pages-Zamora, Vidal, and Brooks. (*Source:* [10], © 2002 IEEE. Reprinted with permission.)

$$\mathbf{r} = \begin{bmatrix} x_T \\ y_T \end{bmatrix} = \begin{bmatrix} x_i \\ y_i \end{bmatrix} + d_i \begin{bmatrix} \cos\phi_i \\ \sin\phi i \end{bmatrix}$$

and, changing these equations from matrix form to conventional form gives:

$$\begin{aligned} x_T &= x_i + d_i \cos\phi_i \\ y_T &= y_i + d_i \sin\phi_i \end{aligned} \tag{2.86}$$

From (2.86),

$$x_T \sin\phi_i = x_i \sin\phi_i + d_i \sin\phi_i \cos\phi_i \tag{2.87}$$

$$y_T \cos\phi_i = y_i \cos\phi_i + d_i \sin\phi_i \cos\phi_i \tag{2.88}$$

Subtract (2.87) from (2.88) and we get

$$-x_i \sin\phi_i + y_i \cos\phi_i = -x_T \sin\phi_i + y_T \cos\phi_i \tag{2.89}$$

Doing this for all the sensors yields, in matrix form,

$$\begin{bmatrix} -x_1 \sin\phi_1 + y_1 \cos\phi_1 \\ -x_2 \sin\phi_2 + y_2 \cos\phi_2 \\ \vdots \\ -x_N \sin\phi_N + y_N \cos\phi_N \end{bmatrix} = \begin{bmatrix} -\sin\phi_1 & \cos\phi_1 \\ -\sin\phi_2 & \cos\phi_2 \\ \vdots & \vdots \\ = \sin\phi_N & \cos\phi_N \end{bmatrix} \begin{bmatrix} x_\mathrm{T} \\ y_\mathrm{T} \end{bmatrix} \tag{2.90}$$

or

$$\mathbf{a}(\boldsymbol{\varphi}) = \mathbf{H}(\boldsymbol{\varphi})\mathbf{x} \tag{2.91}$$

Generally, this matrix expression is over determined—that is, $N > 2$, so there are more equations than unknowns. Therefore, **H** cannot be directly inverted and the pseudoinverse must be used which results from the least-squared solution. The least-squared solution of (2.91) is given by

$$\hat{\mathbf{x}} = \left[\mathbf{H}^\mathrm{T}(\boldsymbol{\varphi})\mathbf{H}(\boldsymbol{\varphi})\right]^{-1} \mathbf{H}^\mathrm{T}(\boldsymbol{\varphi})\mathbf{a}(\boldsymbol{\varphi}) \tag{2.92}$$

Now assume that the bearing measurements contain a small amount of error denoted by $\delta\boldsymbol{\varphi}$ so that $\boldsymbol{\varphi}' = \boldsymbol{\varphi} + \delta\boldsymbol{\varphi}$. It is assumed that $\delta\boldsymbol{\varphi} \sim \mathcal{N}(\mathbf{0}, \boldsymbol{\sigma}^2)$. It is further assumed that the σ_i^2 are uncorrelated with each other. Now,

$$\mathbf{a}(\boldsymbol{\varphi}') = \mathbf{a} + \delta\mathbf{a}(\boldsymbol{\varphi}) \tag{2.93}$$

and

$$\mathbf{H}(\boldsymbol{\varphi}') = \mathbf{H}(\boldsymbol{\varphi}) + \delta\mathbf{H}(\boldsymbol{\varphi}) \tag{2.94}$$

$$\mathbf{a}(\boldsymbol{\varphi}) = \begin{bmatrix} -x_1 \sin\phi_1 + y_1 \cos\phi_1 \\ -x_2 \sin\phi_2 + y_2 \cos\phi_2 \\ \vdots \\ -x_N \sin\phi_N + y_N \cos\phi_N \end{bmatrix} \tag{2.95}$$

so that

$$\frac{\partial \mathbf{a}(\boldsymbol{\varphi})}{\partial \boldsymbol{\varphi}} = \begin{bmatrix} -x_1 \cos\phi_1 - y_1 \sin\phi_1 \\ -x_2 \cos\phi_2 + y_2 \sin\phi_2 \\ \vdots \\ -x_N \cos\phi_N + y_N \sin\phi_N \end{bmatrix} \tag{2.96}$$

Using the approximations that $\delta\boldsymbol{\varphi} \approx \partial\boldsymbol{\varphi}$ and $\delta\mathbf{a} \approx \partial\mathbf{a}$, then

$$\delta\mathbf{a} = \begin{bmatrix} -x_1\delta\phi_1 \cos\phi_1 - y_1\delta\phi_1 \sin\phi_1 \\ -x_2\delta\phi_2 \cos\phi_2 + y_2\delta\phi_2 \sin\phi_2 \\ \vdots \\ -x_N\delta\phi_N \cos\phi_N + y_N\delta\phi_N \sin\phi_N \end{bmatrix} \tag{2.97}$$

Likewise,

$$\mathbf{H}(\boldsymbol{\varphi}) = \begin{bmatrix} -\sin\phi_1 & \cos\phi_1 \\ -\sin\phi_2 & \cos\phi_2 \\ \vdots & \vdots \\ = \sin\phi_N & \cos\phi_N \end{bmatrix} \tag{2.98}$$

so that

$$\frac{\partial \mathbf{H}(\boldsymbol{\varphi})}{\partial \boldsymbol{\varphi}} = \begin{bmatrix} -\cos\phi_1 & -\sin\phi_1 \\ -\cos\phi_2 & -\sin\phi_2 \\ \vdots & \vdots \\ -\cos\phi_N & -\sin\phi_N \end{bmatrix} \tag{2.99}$$

Using the same approximations that lead to (2.97) yields

$$\delta\mathbf{H} = \begin{bmatrix} -\delta\phi_1 \cos\phi_1 & -\delta\phi_1 \sin\phi_1 \\ -\delta\phi_2 \cos\phi_2 & -\delta\phi_2 \sin\phi_2 \\ \vdots & \vdots \\ -\delta\phi_N \cos\phi_N & -\delta\phi_N \sin\phi_N \end{bmatrix} \tag{2.100}$$

Substituting (2.93) and (2.94) into (2.91) yields

$$(\mathbf{H}+\delta\mathbf{H})^{\mathrm{T}}(\mathbf{a}+\delta\mathbf{a})=(\mathbf{H}+\delta\mathbf{H})^{\mathrm{T}}(\mathbf{H}+\delta\mathbf{H})(\mathbf{x}+\delta\mathbf{x}) \tag{2.101}$$

In this expression, $\delta\mathbf{x}$ is identified as the bias in the PF algorithm. Carrying out the multiplications and discarding all terms where two or more incremental terms are products yields

$$\delta\mathbf{x}=(\mathbf{H}^{\mathrm{T}}\mathbf{H})^{-1}\mathbf{H}^{\mathrm{T}}(\delta\mathbf{a}-\delta\mathbf{H}\mathbf{x}) \tag{2.102}$$

Substituting (2.97) and (2.100) into (2.102) shows that $\delta\mathbf{x}=0$ and the estimator is therefore unbiased.

The covariance matrix of $\delta\mathbf{x}$ is given by

$$\mathbf{Q}=\mathcal{E}\{\delta\mathbf{x}\delta\mathbf{x}^{\mathrm{T}}\}=(\mathbf{H}^{\mathrm{T}}\mathbf{H})^{-1}\mathbf{H}^{\mathrm{T}}\mathbf{\Lambda}[(\mathbf{H}^{\mathrm{T}}\mathbf{H})^{-1}\mathbf{H}^{\mathrm{T}}]^{\mathrm{T}} \tag{2.103}$$

where

$$\mathbf{\Lambda}=\begin{bmatrix}\sigma_1^2 d_1^2 & & & \\ & \sigma_2^2 d_2^2 & & \\ & & \ddots & \\ & & & \sigma_N^2 d_N^2\end{bmatrix} \tag{2.104}$$

2.4.4 Total Least-Squared

The method of *total least-squared* (TLS) starts with squaring (2.68):

$$\cos^2(\phi_i)=\frac{(\lambda_{\mathrm{T}}-\lambda_i)^2}{(\lambda_{\mathrm{T}}-\lambda_i)^2+(\psi_{\mathrm{T}}-\psi_i)^2} \tag{2.105}$$

Manipulating this equation eventually yields

$$a_i\lambda_{\mathrm{T}}+b_i\phi_{\mathrm{T}}+c_i\lambda_{\mathrm{T}}^2+d_i\psi_{\mathrm{T}}^2=-e_i \tag{2.106}$$

where

$$a_i=-2\lambda_i\cos^2(\phi_i)+2\lambda_i$$
$$b_i=-2\cos^2(\phi_i)\psi_i\cos^2\lambda_i$$
$$c_i=\cos^2(\phi_i)-1$$

$$d_i = \cos^2(\phi_i)\cos^2\lambda_i$$
$$e_i = \lambda_i^2\cos^2(\phi_i) + \cos^2(\phi_i)\psi_i^2\cos^2\lambda_i - \lambda_i^2$$

For M observations, (2.106) generates M over determined equations as

$$\mathbf{AX} = -\mathbf{E} \tag{2.107}$$

where

$$\mathbf{A} = \begin{bmatrix} a_1 & b_1 & c_1 & d_1 \\ a_2 & b_2 & c_2 & d_2 \\ \vdots & \vdots & \vdots & \vdots \\ a_M & b_M & c_M & d_M \end{bmatrix} \tag{2.108}$$

$$\mathbf{X} = \begin{bmatrix} \lambda_T & \psi_T & \lambda_T^2 & \psi_T^2 \end{bmatrix}^T \tag{2.109}$$

$$\mathbf{E} = \begin{bmatrix} e_1 & e_2 & \cdots & e_M \end{bmatrix}^T \tag{2.110}$$

Equation (2.107) can be rewritten as

$$\mathbf{AX} + \mathbf{E} = \mathbf{0} \tag{2.111}$$

which can be expressed as

$$[\mathbf{A} : -\mathbf{E}][\mathbf{X}^T : -1]^T = \mathbf{0} \tag{2.112}$$

The SVD of the first term in (2.112) can be expressed as

$$[\mathbf{A} : -\mathbf{E}] = \mathbf{USV}^H \triangleq [\mathbf{U}_s \quad \mathbf{U}_o] \begin{bmatrix} \mathbf{S}_s & 0 \\ 0 & 0 \end{bmatrix} \begin{bmatrix} \mathbf{V}_s^H \\ \mathbf{V}_o^H \end{bmatrix} \tag{2.113}$$

The σ_i in **S** are singular values of [**A**: –**E**]. The solution of (2.112) is given by the last column of **V** if the smallest singular value of **A** is greater than the smallest singular value of [**A**:–**E**]. The TLS solution is obtained by scaling $\mathbf{v}_{M+1}$ until the last component is –1:

$$\begin{bmatrix} \mathbf{X}^T & -1 \end{bmatrix} = -\frac{\mathbf{v}_{M+1}}{v_{M+1,M+1}} \tag{2.114}$$

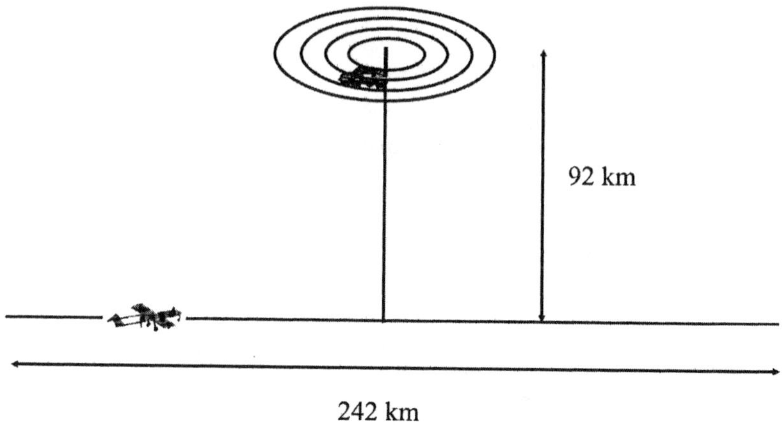

Figure 2.13 Geometry for the TLS example. (*Source:* [11], © 1994 IEEE. Reprinted with permission.)

It is tacitly assumed here that the errors in the bearings are i.i.d. with zero mean.

Example: Rao and Reddy conducted a simulated experiment to evaluate the performance of the TLS approach to geolocation [11]. The geometry studied is illustrated in Figure 2.13, where 122 bearings were taken along the sensor path as shown. The PF was computed using the TLS algorithm as presented. The results for three cases are given in Table 2.1. The estimates produced by the TLS method increased substantially when there was systematic bias in the bearing measurements.

The effects of the SNR on the TLS method were also examined during the same experiment. The results are shown in Figure 2.14. As expected, the error decreased as the SNR increased, with the error dropping to less than 1% of range (from the perpendicular bisector of the baseline) for values of SNR above about 25 dB.

The effects of decreasing the number of observations were also examined for the TLS algorithm, with the results as indicated in Figure 2.15. One percent of

Table 2.1 Results of the TLS Algorithm for the Example

Case	Random error	Systematic error	Error (km)
1	$\sim\mathcal{N}(0, 1)$	0°	0.2637
2	$\sim\mathcal{N}(0, 1)$	–3°	2.917
3	$\sim\mathcal{N}(0, 1)$	6°	7.339

Source: [11].

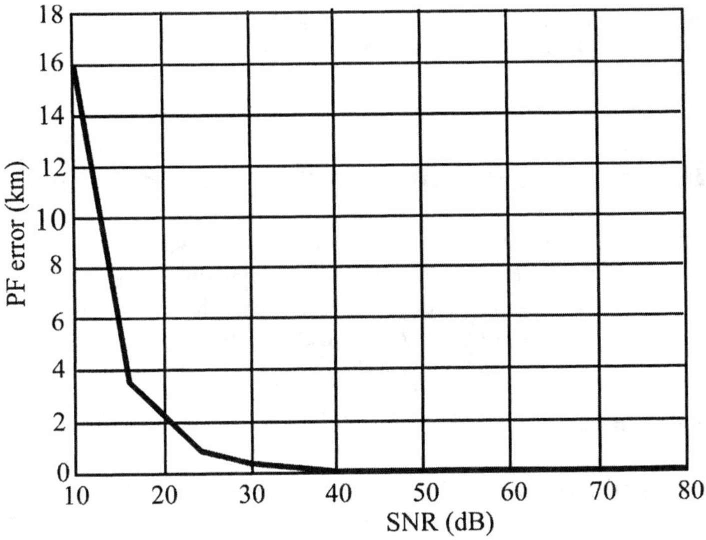

Figure 2.14 PF performance of TLS versus SNR. (*Source:* [11], © 1994 IEEE. Reprinted with permission.)

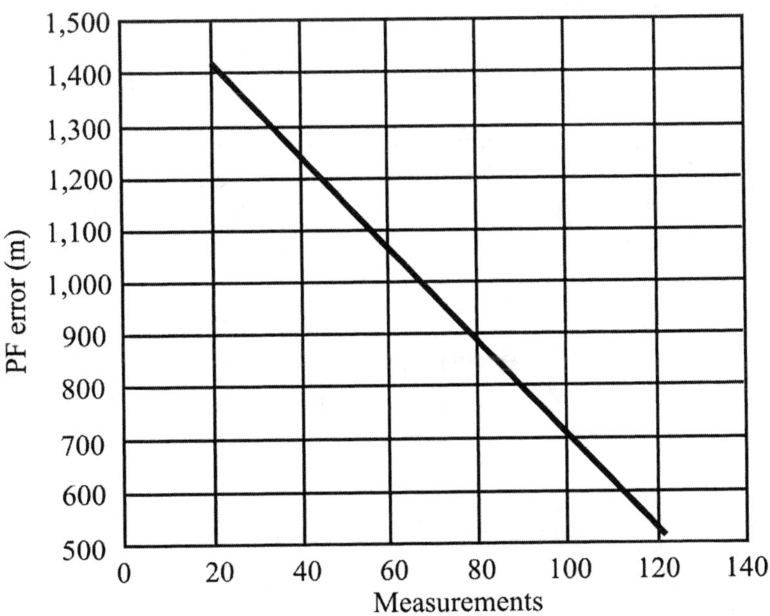

Figure 2.15 TLS performance versus the number of observations. (*Source:* [11], © 1994 IEEE. Reprinted with permission.)

range was achieved for 80 observations and above, but even with only 10 observations the error was only 1.5 km.

■

2.5 Minimum Mean-Squared Error Estimation

PF algorithms are presented in this section that are based on dynamic system models. These algorithms are typically applicable when the measurements are obtained recursively with the PF updated during each iteration. The section begins with a background discussion on dynamical systems before the algorithms are presented.

2.5.1 Dynamical Systems

Dynamical systems can be described by the state transition equation:

$$\mathbf{s}_i = \mathbf{\Phi}_i \mathbf{s}_{i-1} + \mathbf{B}\mathbf{u}_{i-1} + \mathbf{n}_{i-1}, \quad i = 0, 1, \cdots \tag{2.115}$$

with $\mathbf{s} \in \Re^N$ and the measurement vector given by

$$\mathbf{z}_i = \mathbf{H}\mathbf{s}_i + \boldsymbol{\eta}_i, \qquad i = 0, 1, \cdots \tag{2.116}$$

with $\mathbf{z}_i \in \Re^P$, where $\mathbf{s}_i$ is the state of the system at time i and $\mathbf{\Phi}_i$ is the state transition matrix, which can be linear, nonlinear, time variant, or time invariant. Matrix $\mathbf{B}$ relates the effects of the input $\mathbf{u}_{i-1}$ to the state $\mathbf{s}_i$ at sample time i. The vector of random noise $\mathbf{n}_i$ is usually modeled as white noise:

$$\mathcal{E}\{\mathbf{n}_i\} = \mathbf{0} \tag{2.117}$$

$$\mathcal{E}\{\mathbf{n}_i \mathbf{n}_i^{\mathrm{T}}\} = \begin{cases} \sigma_{\mathrm{n}}^2, & i = j \\ 0, & \text{otherwise} \end{cases} \tag{2.118}$$

so

$$\mathbf{Q}_i = \mathrm{diag}[\sigma_{\mathrm{n}}^2 \quad \sigma_{\mathrm{n}}^2 \quad \cdots \quad \sigma_{\mathrm{n}}^2] \tag{2.119}$$

where $\mathbf{Q}_i$ is the covarance matrix of the process noise.

The measurement noise is characterized by

$$\mathcal{E}\{\boldsymbol{\eta}_i\} = \mathbf{0} \tag{2.120}$$

$$\mathcal{E}\{\boldsymbol{\eta}_i\boldsymbol{\eta}_j^{\mathrm{T}}\} = \begin{cases} \sigma_{\eta_i}^2, & i = j \\ 0, & \text{otherwise} \end{cases} \tag{2.121}$$

so

$$\mathbf{R}_i = \mathrm{diag}[\sigma_{\eta_1}^2 \quad \sigma_{\eta_2}^2 \quad \cdots \quad \sigma_{\eta_P}^2] \tag{2.122}$$

is the measurement error covariance matrix. The state variable noise and measurement noise levels are assumed to be independent:

$$\mathcal{E}\{\boldsymbol{\eta}_i\mathbf{n}_j^{\mathrm{T}}\} = 0, \qquad \forall i, j \tag{2.123}$$

The $N \times N$ matrix $\boldsymbol{\Phi}_i$ relates the state at the previous time step $i - 1$ to the state at the current step i, in the absense of any driving function $\mathbf{u}_{i-1}$ or process noise $\mathbf{n}_{i-1}$. The $N \times L$ matrix $\mathbf{B}$ relates the control input $\mathbf{u}$ (if present) to the state $\mathbf{s}_i$. The $P \times N$ matrix $\mathbf{H}$ relates the state to the measurement $\mathbf{z}_i$.

Thus, we wish to determine the values of $\mathbf{s}_i$ at instant i but are not able to measure the states directly. Instead, we measure $\mathbf{z}_i = \tilde{\mathbf{z}}_i$, which are functions of $\mathbf{s}_i$. Knowing these $\mathbf{z}_i$, we wish to estimate the values of $\mathbf{s}_i$. For *minimum mean-squared error* (MMSE) estimation, the optimality criterion used is to minimize the mean of the squared error given by

$$\mathcal{E}\{(\mathbf{s}_i - \hat{\mathbf{s}}_i)^2\} \tag{2.124}$$

We assume that the *probability density functions* (pdf) $p(\mathbf{s}_i)$ and $p(\mathbf{s}_i | \mathbf{z}_i)$ are known. Therefore, we want to find the best guess of $\mathbf{s}_i$ given that $\mathbf{z}_i = \tilde{\mathbf{z}}_i$ in the MMSE sense. In other words, we want to find a function $\hat{\mathbf{s}}_i = g(\hat{\mathbf{z}}_i)$ such that we can minimize

$$J = \mathrm{MSE} = \mathcal{E}\{[\mathbf{s}_i - g(\tilde{\mathbf{z}}_i)]^2 | \mathbf{z}_i = \tilde{\mathbf{z}}_i\} \tag{2.125}$$

Now,

$$\frac{dJ}{dg} = \frac{d}{dg}\mathcal{E}\{[\mathbf{s}_i - g(\tilde{\mathbf{z}}_i)]^2 | \mathbf{z}_i = \tilde{\mathbf{z}}_i\} = 0 \tag{2.126}$$

yields

$$\mathcal{E}\left\{\frac{d}{dg}[\mathbf{s}_i - g(\tilde{\mathbf{z}}_i)]^2\right\} = 0$$

$$\mathcal{E}\left\{\frac{d}{dg}[\mathbf{s}_i^2 - 2\mathbf{s}_i g(\tilde{\mathbf{z}}_i) + g^2(\tilde{\mathbf{z}}_i)]\right\} = 0$$

$$-2\mathcal{E}\{\mathbf{s}_i\} + 2g(\tilde{\mathbf{z}}_i) = 0$$

$$g(\tilde{\mathbf{z}}_i) = \mathcal{E}\{\mathbf{s}_i | \mathbf{z}_i = \tilde{\mathbf{z}}_i\} \qquad (2.127)$$

So the optimum estimate given $\mathbf{z}_i = \tilde{\mathbf{z}}_i$ is the expected value, or mean value, of $\mathbf{s}_i$.

The stochastic function $g(\mathbf{z}_i) = \mathcal{E}\{\mathbf{s}_i | \mathbf{z}_i\}$ minimizes the unconditional *mean-squared error* (MSE) $\mathcal{E}\{[\mathbf{s}_i - g(\mathbf{z}_i)]^2\}$ in addition to minimizing the conditional MSE $\mathcal{E}\{[\mathbf{s}_i - g(\mathbf{z}_i)]^2 | \mathbf{z}_i\}$. Using iterated expectations,

$$\mathcal{E}\{[\mathbf{s}_i - g(\mathbf{z}_i)]^2\} = \mathcal{E}\left\{\mathcal{E}\{[\mathbf{s}_i - g(\mathbf{z}_i)]^2 | \mathbf{z}_i\}\right\}$$

$$= \int_{-\infty}^{\infty} \left\{\mathcal{E}\{[\mathbf{s}_i - g(\tilde{\mathbf{z}}_i)]^2 | \mathbf{z}_i = \tilde{\mathbf{z}}_i\}\right\} p_{\mathbf{z}_i}(\tilde{\mathbf{z}}_i) d\tilde{\mathbf{z}}_i$$

Since $g(\mathbf{z}_i) = \mathcal{E}\{\mathbf{s}_i | \mathbf{z}_i = \tilde{\mathbf{z}}_i\}$ minimizes $\mathcal{E}\{[\mathbf{s}_i - g(\mathbf{z}_i)]^2 | \mathbf{z}_i = \tilde{\mathbf{z}}_i\}$ for each $\tilde{\mathbf{z}}_i$, the integral is also minimized. Let $\hat{\mathbf{s}}_i = \mathcal{E}\{\mathbf{s}_i | \mathbf{z}_i\}$ and $\widehat{\mathbf{s}} = \mathbf{s}_i - \hat{\mathbf{s}}_i$. Then

$$\mathcal{E}\{\widehat{\mathbf{s}}_i | \mathbf{z}_i = \tilde{\mathbf{z}}_i\} = \mathcal{E}\{\mathbf{s}_i - \hat{\mathbf{s}}_i | \mathbf{z}_i = \tilde{\mathbf{z}}_i\}$$

$$= \mathcal{E}\{\mathbf{s}_i | \mathbf{z}_i = \tilde{\mathbf{z}}_i\} - \mathcal{E}\{\hat{\mathbf{s}}_i | \mathbf{z}_i = \tilde{\mathbf{z}}_i\}$$

$$= \mathcal{E}\{\mathbf{s}_i | \mathbf{z}_i = \tilde{\mathbf{z}}_i\} - \mathcal{E}\{\mathbf{s}_i | \mathbf{z}_i = \tilde{\mathbf{z}}_i\}$$

$$= 0$$

$$\mathcal{E}\{\widehat{\mathbf{s}}_i\} = \mathcal{E}\left\{\mathcal{E}\{\mathbf{s}_i | \mathbf{z}_i\}\right\} = 0$$

so the MMSE estimator is unbiased.

In addition,

$$\mathbf{Cov}(\widehat{\mathbf{s}}_i, \hat{\mathbf{s}}_i) = \mathcal{E}\left\{(\widehat{\mathbf{s}}_i - \mathcal{E}\{\widehat{\mathbf{s}}_i\})(\hat{\mathbf{s}}_i - \mathcal{E}\{\hat{\mathbf{s}}_i\})\right\}$$

$$
\begin{aligned}
&= \mathcal{E}\left\{\widehat{\mathbf{s}}_i(\hat{\mathbf{s}}_i - \mathcal{E}\{\mathbf{s}_i\})\right\} \\
&= \mathcal{E}\left\{\mathcal{E}\{\widehat{\mathbf{s}}_i(\hat{\mathbf{s}}_i - \mathcal{E}\{\mathbf{s}_i\}) | \mathbf{z}_i\}\right\} \\
&= \mathcal{E}\left\{(\hat{\mathbf{s}}_i - \mathcal{E}\{\mathbf{s}_i\}) \underbrace{\mathcal{E}\{\widehat{\mathbf{s}}_i | \mathbf{z}_i\}}_{=0}\right\} = 0
\end{aligned}
$$

so $\widehat{\mathbf{s}}_i$ and $\hat{\mathbf{s}}_i$ are uncorrelated, and as a result,

$$
\mathbf{Var}(\mathbf{s}_i) = \mathbf{Var}(\hat{\mathbf{s}}_i) + \mathbf{Var}(\widehat{\mathbf{s}}_i) \tag{2.128}
$$

2.5.2 Linear Minimum Mean-Squared Estimation

The MMSE estimator requires the a posterior density $p(\boldsymbol{\theta}|\mathbf{x})$ ($p(\mathbf{s}_i|\mathbf{z}_i)$ above) to be known, which is frequently difficult to calculate. If the analysis is restricted to the class of estimators that are affine (linear with a possible translation) functions of the observation vector **x**, however, the *linear minimum mean-squared error* (LMMSE) estimator ensues which is a function of the first and second moments of $\boldsymbol{\theta}$ and **x** only. Furthermore, if **x** and $\boldsymbol{\theta}$ are jointly Gaussian, the LMMSE estimator is the same as the MMSE estimator.

The LMMSE estimator for an M-dimensional unknown vector $\boldsymbol{\theta}$ and an N-dimensional observation vector **x** will be considered, which are of the form

$$
\hat{\boldsymbol{\theta}} = \mathbf{H}\mathbf{x} + \mathbf{b} \tag{2.129}
$$

where **H** is an $M \times N$ matrix and **b** is an M element vector. **H** and **b** are chosen to minimize

$$
\mathcal{E}_{x,\theta}\left\{\left(\boldsymbol{\theta} - \hat{\boldsymbol{\theta}}\right)^{\mathrm{T}}\left(\boldsymbol{\theta} - \hat{\boldsymbol{\theta}}\right)\right\} \tag{2.130}
$$

The LMMSE estimate is

$$
\hat{\boldsymbol{\theta}} = \mathcal{E}\{\boldsymbol{\theta}\} + \mathbf{R}_{\theta x}\mathbf{R}_{xx}^{-1}\left(\mathbf{x} - \mathcal{E}\{\mathbf{x}\}\right) \tag{2.131}
$$

The value $\mathcal{E}\{\boldsymbol{\theta}\}$ is the mean of $\boldsymbol{\theta}$, $\mathcal{E}\{\mathbf{x}\}$ is the mean of **x**, $\mathbf{R}_{\theta\mathrm{x}}$ is the cross covariance matrix of $\boldsymbol{\theta}$ and **x** given by

$$\mathbf{R}_{\theta x} = \mathcal{E}\left\{(\boldsymbol{\theta} - \mathcal{E}\{\boldsymbol{\theta}\})(\mathbf{x} - \mathcal{E}\{\mathbf{x}\})^{\mathrm{T}}\right\} \tag{2.132}$$

and $\mathbf{R}_{xx}$ is the autocovariance matrix of $\mathbf{x}$ given by

$$\mathbf{R}_{xx} = \mathcal{E}\left\{(\mathbf{x} - \mathcal{E}\{\mathbf{x}\})(\mathbf{x} - \mathcal{E}\{\mathbf{x}\})^{\mathrm{T}}\right\} \tag{2.133}$$

Assume that $\mathbf{x}$ and $\boldsymbol{\theta}$ have means of zero. Then the orthogonality principle applies.

Property: Orthogonality Principle

For the LMMSE estimate $\hat{\boldsymbol{\theta}}$, the estimate error is orthogonal to the data. That is,

$$\mathcal{E}\left\{(\boldsymbol{\theta} - \hat{\boldsymbol{\theta}})\mathbf{x}^{\mathrm{T}}\right\} = \mathbf{0} \tag{2.134}$$

Note that $\mathbf{0}$ is an $M \times N$ matrix.

Proof: Let $\mathbf{A}$ be an $M \times N$ matrix for which the orthogonality principle holds:

$$\mathcal{E}\left\{(\boldsymbol{\theta} - \mathbf{A}\mathbf{x})\mathbf{x}^{\mathrm{T}}\right\} = \mathbf{0} \tag{2.135}$$

Let $\mathbf{B}$ be any other $M \times N$ matrix and $\mathbf{Bx}$ be the estimate that it gives for $\boldsymbol{\theta}$. The expected squared error of this estimate is

$$\begin{aligned}
\mathcal{E}\left\{(\boldsymbol{\theta} - \mathbf{Bx})^{\mathrm{T}}(\boldsymbol{\theta} - \mathbf{Bx})\right\} &= \mathcal{E}\left\{(\boldsymbol{\theta} - \mathbf{Ax} + \mathbf{Ax} - \mathbf{Bx})^{\mathrm{T}}(\boldsymbol{\theta} - \mathbf{Ax} + \mathbf{Ax} - \mathbf{Bx})\right\} \\
&= \mathcal{E}\left\{[\boldsymbol{\theta} - \mathbf{Ax} + (\mathbf{A} - \mathbf{B})\mathbf{x}]^{\mathrm{T}}[\boldsymbol{\theta} - \mathbf{Ax} + (\mathbf{A} - \mathbf{B})\mathbf{x}]\right\} \\
&= \mathcal{E}\left\{(\boldsymbol{\theta} - \mathbf{Ax})^{\mathrm{T}}(\boldsymbol{\theta} - \mathbf{Ax})\right\} \\
&\quad + \mathcal{E}\left\{(\boldsymbol{\theta} - \mathbf{Ax})^{\mathrm{T}}(\mathbf{A} - \mathbf{B})\mathbf{x}\right\} \\
&\quad + \mathcal{E}\left\{[(\mathbf{A} - \mathbf{B})\mathbf{x}]^{\mathrm{T}}(\boldsymbol{\theta} - \mathbf{Ax})\right\} \\
&\quad + \mathcal{E}\left\{[(\mathbf{A} - \mathbf{B})\mathbf{x}]^{\mathrm{T}}[(\mathbf{A} - \mathbf{B})\mathbf{x}]\right\}
\end{aligned} \tag{2.136}$$

$$
\begin{aligned}
\mathcal{E}\left\{(\boldsymbol{\theta}-\mathbf{A}\mathbf{x})^{\mathrm{T}}(\mathbf{A}-\mathbf{B})\mathbf{x}\right\} &= \operatorname{tr}\mathcal{E}\left\{(\boldsymbol{\theta}-\mathbf{A}\mathbf{x})\left[(\mathbf{A}-\mathbf{B})\mathbf{x}\right]^{\mathrm{T}}\right\} \\
&= \operatorname{tr}\mathcal{E}\left\{(\boldsymbol{\theta}-\mathbf{A}\mathbf{x})\mathbf{x}^{\mathrm{T}}(\mathbf{A}-\mathbf{B})^{\mathrm{T}}\right\} \\
&= \operatorname{tr}\mathcal{E}\left\{(\boldsymbol{\theta}-\mathbf{A}\mathbf{x})\mathbf{x}^{\mathrm{T}}\right\}(\mathbf{A}-\mathbf{B})^{\mathrm{T}} \\
&= \operatorname{tr}\mathbf{0}(\mathbf{A}-\mathbf{B})^{\mathrm{T}} \\
&= 0
\end{aligned} \tag{2.137}
$$

In a similar fashion, it can be shown that $\mathcal{E}\left\{\left[(\mathbf{A}-\mathbf{B})\mathbf{x}\right]^{\mathrm{T}}(\boldsymbol{\theta}-\mathbf{A}\mathbf{x})\right\}=\mathbf{0}$. Thus, the squared error becomes

$$
\begin{aligned}
\mathcal{E}\left\{(\boldsymbol{\theta}-\mathbf{B}\mathbf{x})^{\mathrm{T}}(\boldsymbol{\theta}-\mathbf{B}\mathbf{x})\right\} &= \mathcal{E}\left\{(\boldsymbol{\theta}-\mathbf{A}\mathbf{x})^{\mathrm{T}}(\boldsymbol{\theta}-\mathbf{A}\mathbf{x})\right\} \\
&\quad + \mathcal{E}\left\{\left[(\mathbf{A}-\mathbf{B})\mathbf{x}\right]^{\mathrm{T}}(\mathbf{A}-\mathbf{B})\mathbf{x}\right\}
\end{aligned} \tag{2.138}
$$

but

$$
\mathcal{E}\left\{\left[(\mathbf{A}-\mathbf{B})\mathbf{x}\right]^{\mathrm{T}}(\mathbf{A}-\mathbf{B})\mathbf{x}\right\} \geq 0 \tag{2.139}
$$

so

$$
\mathcal{E}\left\{(\boldsymbol{\theta}-\mathbf{B}\mathbf{x})^{\mathrm{T}}(\boldsymbol{\theta}-\mathbf{B}\mathbf{x})\right\} \geq \mathcal{E}\left\{(\boldsymbol{\theta}-\mathbf{A}\mathbf{x})^{\mathrm{T}}(\boldsymbol{\theta}-\mathbf{A}\mathbf{x})\right\} \tag{2.140}
$$

Thus, the expected squared error of any linear estimator cannot be smaller than the expected squared error of a linear estimator that satisfies the orthogonality principle. Therefore, the LMMSE estimator must satisfy the orthogonality principle.

■

There is a geometric interpretation of the orthogonality principle. A collection of random variables can be considered as a Hilbert space (a vector space). The random variables x_0 through x_{N-1} and $\boldsymbol{\theta}$ are elements of the space. The inner product between elements is

$$
\left(x_i, x_j\right) = \mathcal{E}\left\{x_i x_j\right\} \tag{2.141}
$$

The magnitude of an element of the space is

$$\|x_i\| = \sqrt{(x_i, x_i)} = \sqrt{\mathcal{E}\{x_i x_i\}} \tag{2.142}$$

The vector $\hat{\boldsymbol{\theta}}$ is a linear combination of x_0 through x_{N-1}. To make the magnitude of the estimate error as small as possible, the estimate is a projection of $\hat{\boldsymbol{\theta}}$ onto x_0 through x_{N-1}; in other words, the error should be perpendicular to x_0 through x_{N-1}.

An example when $N = 2$ is illustrated in Figure 2.16. The projection of $\hat{\boldsymbol{\theta}}$ onto the plane formed by the measurement vectors will be the smallest (zero in this case) when $\hat{\boldsymbol{\theta}}$ is orthogonal to that plane.

Given that the estimator has the form $\hat{\boldsymbol{\theta}} = \mathbf{Hx}$, the orthogonality principle is used to solve for the value of $\mathbf{H}$ as follows.

$$\mathcal{E}\{(\boldsymbol{\theta} - \mathbf{Hx})\mathbf{x}^{\mathrm{T}}\} = \mathbf{0} \tag{2.143}$$

Expanding the inner product,

$$\mathcal{E}\{\boldsymbol{\theta}\mathbf{x}^{\mathrm{T}}\} - \mathcal{E}\{\mathbf{Hxx}^{\mathrm{T}}\} = \mathcal{E}\{\boldsymbol{\theta}\mathbf{x}^{\mathrm{T}}\} - \mathbf{H}\mathcal{E}\{\mathbf{xx}^{\mathrm{T}}\} = \mathbf{0} \tag{2.144}$$

These expected value terms are identified as the cross-covariance of $\boldsymbol{\theta}$ and $\mathbf{x}$ and the autocovariance of $\mathbf{x}$, respectively. Therefore,

$$\mathbf{R}_{\theta x} - \mathbf{HR}_{xx} = \mathbf{0} \tag{2.145}$$

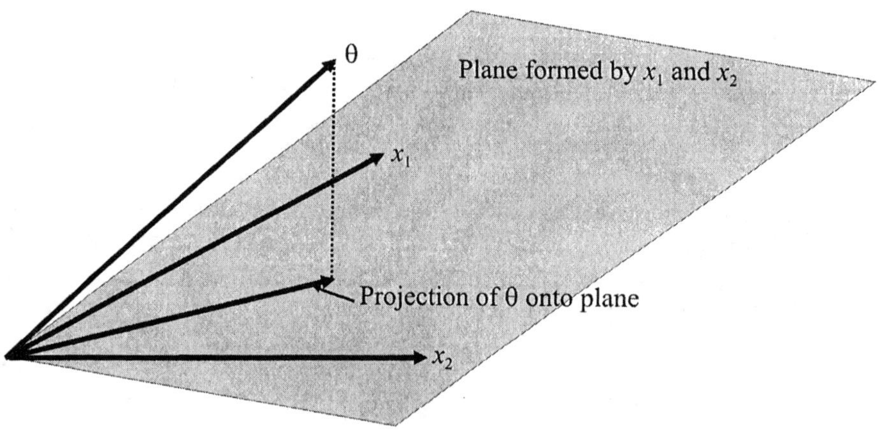

Figure 2.16 $N = 2$ example of projection of $\boldsymbol{\theta}$ onto the plane formed by x_1 and x_2.

and

$$\mathbf{H} = \mathbf{R}_{\theta x}\mathbf{R}_{xx}^{-1} \tag{2.146}$$

so

$$\hat{\boldsymbol{\theta}} = \mathbf{R}_{\theta x}\mathbf{R}_{xx}^{-1}\mathbf{x} \tag{2.147}$$

If $\mathbf{x}$ and $\boldsymbol{\theta}$ do not have zero means, new random variables are created that do have zero means as

$$\tilde{\mathbf{x}} = \mathbf{x} - \mathcal{E}\{\mathbf{x}\} \tag{2.148}$$

$$\tilde{\boldsymbol{\theta}} = \boldsymbol{\theta} - \mathcal{E}\{\boldsymbol{\theta}\} \tag{2.149}$$

The LMMSE estimate of $\tilde{\boldsymbol{\theta}}$ is

$$\begin{aligned}\hat{\tilde{\boldsymbol{\theta}}} &= \mathbf{R}_{\tilde{\theta}\tilde{x}}\mathbf{R}_{\tilde{x}\tilde{x}}^{-1}\tilde{\mathbf{x}} \\ &= \mathbf{R}_{\theta x}\mathbf{R}_{xx}^{-1}\tilde{\mathbf{x}}\end{aligned} \tag{2.150}$$

From (2.149),

$$\hat{\boldsymbol{\theta}} = \hat{\tilde{\boldsymbol{\theta}}} + \mathcal{E}\{\boldsymbol{\theta}\} = \mathcal{E}\{\boldsymbol{\theta}\} + \mathbf{R}_{\theta x}\mathbf{R}_{xx}^{-1}\left(\mathbf{x} - \mathcal{E}\{\mathbf{x}\}\right) \tag{2.151}$$

This estimator is unbiased:

$$\mathcal{E}\{\hat{\boldsymbol{\theta}}\} = \mathcal{E}\{\boldsymbol{\theta}\} + \mathbf{R}_{\theta x}\mathbf{C}_{xx}^{-1}\mathcal{E}\{(\mathbf{x} - \mathcal{E}\{\mathbf{x}\})\} = 0 \tag{2.152}$$

with the expected squared error matrix given by

$$\mathcal{E}\{(\mathbf{q}-\hat{\mathbf{q}})(\mathbf{q}-\hat{\mathbf{q}})\mathbf{x}^{\mathrm{T}}\} = \mathcal{E}\left\{\begin{matrix}[\mathbf{q} - \mathcal{E}\{\mathbf{q}\} - \mathbf{R}_{\theta x}\mathbf{R}_{xx}^{-1}(\mathbf{x} - \mathcal{E}\{\mathbf{x}\})] \\ \times[\mathbf{q} - \mathcal{E}\{\mathbf{q}\} - \mathbf{R}_{\theta x}\mathbf{R}_{xx}^{-1}(\mathbf{x} - \mathcal{E}\{\mathbf{x}\})]^{\mathrm{T}}\end{matrix}\right\}$$

$$
\begin{aligned}
&= \mathcal{E}\left\{(\mathbf{q}-\mathcal{E}\{\mathbf{q}\})(\mathbf{q}-\mathcal{E}\{\mathbf{q}\})^{\mathrm{T}}\right\} \\
&\qquad -\mathcal{E}\left\{(\mathbf{q}-\mathcal{E}\{\mathbf{q}\})\left[\mathbf{R}_{\theta x}\mathbf{R}_{xx}^{-1}(\mathbf{x}-\mathcal{E}\{\mathbf{x}\})^{\mathrm{T}}\right]\right\} \\
&\qquad -\mathcal{E}\left\{\left[\mathbf{R}_{\theta x}\mathbf{R}_{xx}^{-1}(\mathbf{x}-\mathcal{E}\{\mathbf{x}\})\right](\mathbf{q}-\mathcal{E}\{\mathbf{q}\})^{\mathrm{T}}\right\} \\
&\qquad +\mathcal{E}\left\{\left[\mathbf{R}_{\theta x}\mathbf{R}_{xx}^{-1}(\mathbf{x}-\mathcal{E}\{\mathbf{x}\})\right]\left[\mathbf{R}_{\theta x}\mathbf{R}_{xx}^{-1}(\mathbf{x}-\mathcal{E}\{\mathbf{x}\})\right]^{\mathrm{T}}\right\} \\
&= \mathcal{E}\left\{(\mathbf{q}-\mathcal{E}\{\mathbf{q}\})(\mathbf{q}-\mathcal{E}\{\mathbf{q}\})^{\mathrm{T}}\right\} \\
&\qquad -\mathcal{E}\left\{(\mathbf{q}-\mathcal{E}\{\mathbf{q}\})\left[\mathbf{R}_{\theta x}\mathbf{R}_{xx}^{-1}(\mathbf{x}-\mathcal{E}\{\mathbf{x}\})^{\mathrm{T}}\right]\right\} \\
&\qquad -\mathcal{E}\left\{\left[\mathbf{R}_{\theta x}\mathbf{R}_{xx}^{-1}(\mathbf{x}-\mathcal{E}\{\mathbf{x}\})\right](\mathbf{q}-\mathcal{E}\{\mathbf{q}\})^{\mathrm{T}}\right\} \\
&\qquad +\mathcal{E}\left\{\left[\mathbf{R}_{\theta x}\mathbf{R}_{xx}^{-1}(\mathbf{x}-\mathcal{E}\{\mathbf{x}\})\right]\left[\mathbf{R}_{\theta x}\mathbf{R}_{xx}^{-1}(\mathbf{x}-\mathcal{E}\{\mathbf{x}\})\right]^{\mathrm{T}}\right\} \\
&= \mathcal{E}\left\{(\mathbf{q}-\mathcal{E}\{\mathbf{q}\})(\mathbf{q}-\mathcal{E}\{\mathbf{q}\})^{\mathrm{T}}\right\} \\
&\qquad -\mathcal{E}\left\{(\mathbf{q}-\mathcal{E}\{\mathbf{q}\})(\mathbf{x}-\mathcal{E}\{\mathbf{x}\})^{\mathrm{T}}(\mathbf{R}_{xx}^{-1})^{\mathrm{T}}\mathbf{R}_{\theta x}^{\mathrm{T}}\right\} \\
&\qquad -\mathcal{E}\left\{\left[\mathbf{R}_{\theta x}\mathbf{R}_{xx}^{-1}(\mathbf{x}-\mathcal{E}\{\mathbf{x}\})\right](\mathbf{q}-\mathcal{E}\{\mathbf{q}\})^{\mathrm{T}}\right\} \\
&\qquad +\mathcal{E}\left\{\left[\mathbf{R}_{\theta x}\mathbf{R}_{xx}^{-1}(\mathbf{x}-\mathcal{E}\{\mathbf{x}\})(\mathbf{x}-\mathcal{E}\{\mathbf{x}\})^{\mathrm{T}}(\mathbf{R}_{xx}^{-1})^{\mathrm{T}}\mathbf{R}_{\theta x}^{\mathrm{T}}\right]\right\} \\
&= \mathbf{R}_{\theta\theta} - \mathbf{R}_{\theta x}\mathbf{R}_{xx}^{-1}\mathbf{R}_{x\theta} \\
&\qquad -\mathbf{R}_{\theta x}\mathbf{R}_{xx}^{-1}\mathbf{R}_{x\theta} + \mathbf{R}_{\theta x}\mathbf{R}_{xx}^{-1}\mathbf{R}_{xx}\mathbf{R}_{xx}^{-1}\mathbf{R}_{x\theta} \\
&= \mathbf{R}_{\theta\theta} - \mathbf{R}_{\theta x}\mathbf{R}_{xx}^{-1}\mathbf{R}_{x\theta}
\end{aligned}
\tag{2.153}
$$

Example: $N = 1$ and θ has mean μ_θ and variance σ_θ^2:

$$x_0 = \theta + n_0 \tag{2.154}$$

This example might apply to a single LOB from a single sensor. The noise $n_0 \sim \mathcal{N}\left(0, \sigma_N^2\right)$ and is independent of θ.

$$\mu_x = \mathcal{E}\{x_0\} = \mu_\theta \tag{2.155}$$

$$
\begin{aligned}
r_{xx} &= \mathcal{E}\left\{(x_0-\mu_x)^2\right\} \\
&= \mathcal{E}\left\{(\theta+n_0-\mu_\theta)^2\right\} \\
&= \mathcal{E}\left\{(\theta-\mu_\theta)^2\right\}+\mathcal{E}\left\{2(\theta-\mu_\theta)n_0\right\}+\mathcal{E}\left\{n_0^2\right\} \\
&= \sigma_\theta^2+\sigma_N^2
\end{aligned} \tag{2.156}
$$

$$
\begin{aligned}
r_{\theta x} &= \mathcal{E}\left\{(\theta-\mu_\theta)(x_0-\mu_x)\right\} \\
&= \mathcal{E}\left\{(\theta-\mu_\theta)((\theta-\mu_\theta)+n_0)\right\} \\
&= \mathcal{E}\left\{(\theta-\mu_\theta)(\theta-\mu_\theta)\right\}+\mathcal{E}\left\{(\theta-\mu_\theta)n_0\right\} \\
&= \sigma_\theta^2
\end{aligned} \tag{2.157}
$$

$$
\begin{aligned}
\hat{\theta} &= \mu_\theta + r_{\theta x} r_{xx}^{-1}(x_0-\mu_x) \\
&= \mu_\theta + \frac{\sigma_\theta^2}{\sigma_\theta^2+\sigma_N^2}(x_0-\mu_\theta) \\
&= \frac{\sigma_\theta^2}{\sigma_\theta^2+\sigma_N^2}x_0 + \frac{\sigma_N^2}{\sigma_\theta^2+\sigma_N^2}\mu_\theta
\end{aligned} \tag{2.158}
$$

The expected squared error for this estimator is

$$
\begin{aligned}
\mathcal{E}\left\{(\hat{\theta}-\theta)^2\right\} &= r_{\theta\theta} - r_{\theta x} r_{xx}^{-1} r_{x\theta} \\
&= \sigma_\theta^2 - \frac{(\sigma_\theta^2)^2}{\sigma_\theta^2+\sigma_N^2} \\
&= \frac{1}{\frac{1}{\sigma_\theta^2}+\frac{1}{\sigma_N^2}}
\end{aligned} \tag{2.159}
$$

■

2.5.3 Target Location Estimation with the Linear Model

For the linear model, the observation vector is given by **x**, which consists of N random observations. The parameters to be estimated are given by the M element vector $\boldsymbol{\theta}$, with means given by the mean vector $\boldsymbol{\mu}_\theta$ and covariance matrix **R**. The observation matrix, **H**, is known, deterministic, and is $N \times M$. The noise, **n**, is an

N-element vector with mean zero and covariance matrix $\mathbf{R}_{nn}$. It is assumed that $\boldsymbol{\theta}$ and $\mathbf{n}$ are uncorrelated.

The model is given by

$$\mathbf{x} = \mathbf{H}\boldsymbol{\theta} + \mathbf{n} \tag{2.160}$$

As seen in Figure 2.17, the observations consist of the azimuth of the target (ϕ), elevation of the target (φ), and range to the target (r) corrupted by noise:

$$\mathbf{x} = \begin{bmatrix} \phi \\ \varphi \\ r \end{bmatrix} + \mathbf{n} \tag{2.161}$$

The unknown is the target position in Cartesian coordinates:

$$\boldsymbol{\theta} = \begin{bmatrix} \theta_x & \theta_y & \theta_z \end{bmatrix}^{\mathrm{T}} \tag{2.162}$$

where $\theta_x = x_{\mathrm{T}}$, $\theta_y = y_{\mathrm{T}}$, and $\theta_z = z_{\mathrm{T}}$. The relationship between the observations and the target position is

$$\phi = \tan^{-1} \frac{\theta_y - y_i}{\theta_x - x_i} \tag{2.163}$$

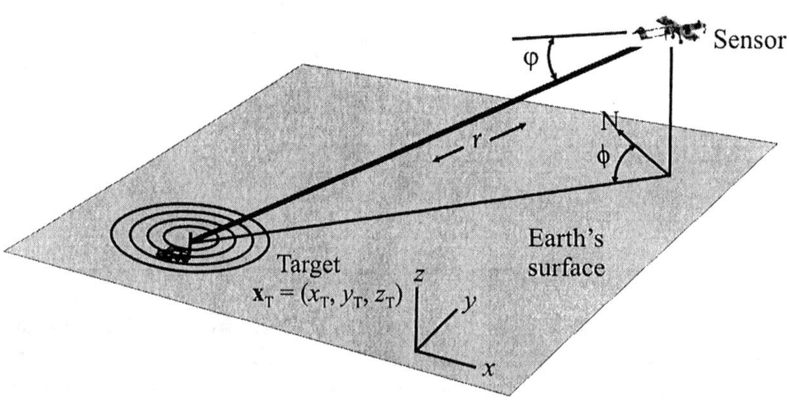

Figure 2.17 Target location geometry.

$$\varphi = \tan^{-1} \frac{\theta_z - z_i}{\sqrt{(\theta_x - x_i)^2 + (\theta_y - y_i)^2}} \tag{2.164}$$

and

$$r = \sqrt{(\theta_x - x_i)^2 + (\theta_y - y_i)^2 + (\theta_z - z_i)^2} \tag{2.165}$$

where the position of the sensor at instant i is given by $\mathbf{x}_s = (x_i, y_i, z_i)$. These are put into matrix form as

$$\mathbf{x} = \mathbf{H}\boldsymbol{\theta} + \mathbf{n} \tag{2.166}$$

Note the squares, square roots, and $\tan^{-1}$ functions in (2.163) to (2.165) are nonlinear functions and therefore do not fit the basic assumptions given above. However, (2.166) can be expanded in a Taylor series about some nominal value $\boldsymbol{\theta}_0$ to get

$$\mathbf{x} \approx \mathbf{H}\boldsymbol{\theta}_0 + \mathbf{h}(\boldsymbol{\theta} - \boldsymbol{\theta}_0) + \mathbf{n} \tag{2.167}$$

where

$$\mathbf{h} = \left. \frac{\partial \mathbf{H}\boldsymbol{\theta}}{\partial \boldsymbol{\theta}} \right|_{\boldsymbol{\theta} = \boldsymbol{\theta}_0} \tag{2.168}$$

One possibility is to set $\boldsymbol{\theta}_0 = \boldsymbol{\mu}_\theta$. Make the following definitions:

$$\tilde{\mathbf{x}} = \mathbf{x} - \mathbf{H}(\boldsymbol{\mu}_\theta) \tag{2.169}$$

$$\tilde{\boldsymbol{\theta}} = \boldsymbol{\theta} - \boldsymbol{\mu}_\theta \tag{2.170}$$

and

$$\tilde{\mathbf{x}} = \mathbf{H}\tilde{\boldsymbol{\theta}} + \mathbf{n} \tag{2.171}$$

This is now in the form of the linear model.

The LMMSE estimator for this linear model is found as follows.

$$\boldsymbol{\mu}_x = \mathcal{E}\{\mathbf{x}\} = \mathcal{E}\{\mathbf{H}\boldsymbol{\theta} + \mathbf{n}\} = \mathbf{H}\boldsymbol{\mu}_\theta \tag{2.172}$$

$$
\begin{aligned}
\mathbf{R}_{xx} &= \mathcal{E}\left\{(\mathbf{x}-\boldsymbol{\mu}_x)(\mathbf{x}-\boldsymbol{\mu}_x)^{\mathrm{T}}\right\} \\
&= \mathcal{E}\left\{(\mathbf{H}\boldsymbol{\theta}+\mathbf{n}-\mathbf{H}\boldsymbol{\mu}_\theta)(\mathbf{H}\boldsymbol{\theta}+\mathbf{n}-\mathbf{H}\boldsymbol{\mu}_\theta)^{\mathrm{T}}\right\} \\
&= \mathbf{H}\mathbf{R}_{\theta\theta}\mathbf{H}^{\mathrm{T}}+\mathbf{R}_{nn} \qquad (2.173)\\
&= \mathcal{E}\left\{\mathbf{H}(\boldsymbol{\theta}-\boldsymbol{\mu}_\theta)(\boldsymbol{\theta}-\boldsymbol{\mu}_\theta)^{\mathrm{T}}\right\}+\mathcal{E}\left\{\mathbf{H}(\boldsymbol{\theta}-\boldsymbol{\mu}_\theta)\mathbf{n}^{\mathrm{T}}\right\} \\
&\qquad +\mathcal{E}\left\{\mathbf{n}(\boldsymbol{\theta}-\boldsymbol{\mu}_\theta)^{\mathrm{T}}\mathbf{H}^{\mathrm{T}}\right\}+\mathcal{E}\left\{\mathbf{n}\mathbf{n}^{\mathrm{T}}\right\} \qquad (2.174)\\
\mathbf{R}_{\theta x} &= \mathcal{E}\left\{(\boldsymbol{\theta}-\boldsymbol{\mu}_\theta)(\mathbf{x}-\boldsymbol{\mu}_x)\right\} \\
&= \mathcal{E}\left\{\boldsymbol{\theta}\boldsymbol{\mu}_\theta(\mathbf{H}\boldsymbol{\theta}+\mathbf{n}-\mathbf{H}\boldsymbol{\mu}_\theta)^{\mathrm{T}}\right\} \\
&= \mathcal{E}\left\{(\boldsymbol{\theta}-\boldsymbol{\mu}_\theta)(\boldsymbol{\theta}-\boldsymbol{\mu}_\theta)^{\mathrm{T}}\mathbf{H}^{\mathrm{T}}\right\}+\mathcal{E}\left\{(\boldsymbol{\theta}-\boldsymbol{\mu}_\theta)\mathbf{n}^{\mathrm{T}}\right\} \\
&= \mathbf{R}_{\theta\theta}\mathbf{H}^{\mathrm{T}} \qquad (2.175)
\end{aligned}
$$

and

$$
\begin{aligned}
\hat{\boldsymbol{\theta}} &= \boldsymbol{\mu}_\theta+\mathbf{R}_{\theta x}\mathbf{R}_{xx}^{-1}(\mathbf{x}-\boldsymbol{\mu}_x) \\
&= \boldsymbol{\mu}_\theta+\mathbf{R}_{\theta\theta}\mathbf{H}^{\mathrm{T}}\left(\mathbf{H}\mathbf{R}_{\theta\theta}\mathbf{H}^{\mathrm{T}}+\mathbf{R}_{nn}\right)^{-1}(\mathbf{x}-\mathbf{H}\boldsymbol{\mu}_\theta) \qquad (2.176)
\end{aligned}
$$

The squared error matrix from (2.153) is

$$
\begin{aligned}
\mathcal{E}\left\{(\hat{\boldsymbol{\theta}}-\boldsymbol{\theta})(\hat{\boldsymbol{\theta}}-\boldsymbol{\theta})^{\mathrm{T}}\right\} &= \mathbf{R}_{\theta\theta}-\mathbf{R}_{\theta x}\mathbf{R}_{xx}^{-1}\mathbf{R}_{x\theta} \\
&= \mathbf{R}_{\theta\theta}-\mathbf{R}_{\theta\theta}\mathbf{H}^{\mathrm{T}}\left(\mathbf{H}\mathbf{R}_{\theta\theta}\mathbf{H}^{\mathrm{T}}+\mathbf{R}_{nn}\right)^{-1}\mathbf{H}\mathbf{R}_{\theta\theta} \\
&= \left(\mathbf{R}_{\theta\theta}^{-1}+\mathbf{H}^{\mathrm{T}}\mathbf{R}_{nn}^{-1}\mathbf{H}\right)^{-1} \qquad (2.177)
\end{aligned}
$$

2.5.4 Kalman Filter Methods

In some cases, knowing only the PF of a target is inadequate. Examples of this include subsurface oceanographic targets that are tracked with active sonar and surface and airborne targets that are tracked by active radar. Thus, methods were devised to accommodate such tracking. Kalman filtering is one of those methods, although Kalman filters have much broader applications than tracking targets. Kalman filtering is based on the principles of MMSE.

In 1960, R. E. Kalman published a seminal paper describing a recursive solution to the discrete-data linear filtering problem [12]. That solution was based on modeling dynamic linear systems as discrete-data systems, and the resulting state variables are used to solve the Wiener filtering problem. Kalman filtering is a technique for optimally estimating the state of a dynamical system at time t_1 given the value of the state variables at time t. If $t_1 < t$, then the estimate is an *interpolation* of the state variables. If $t_1 = t$, then the system is a *filter*. If $t_1 > t$, the problem is one of *prediction* of a future state given the curent state. Of concern here is the filtering problem.

Kalman filter estimation uses a form of feedback control: the filter estimates the process state at some time and then obtains feedback in the form of noisy measurements. The behavior of a dynamic system can be described by the time evolution of the state variables. As indicated previously, the state variables of a dynamic system cannot be determined exactly by direct measurements; instead, the measurements available are functions of the state variables corrupted by random noise. It is then necessary to estimate the state variables from the noisy observations. The measurement and estimation process is illustrated in Figure 2.18. The purpose of the Kalman filter is to optimally estimate the state variables in the dynamical system. *Optimum* in *this* sense means that the mean squared estimation error is minimized.

The Kalman filter can be applied to the PF estimation problem in a variety of ways, two of which are presented here. In the case of the standard Kalman filter, the results are optimum in the least mean-squared error sense. In the case of the

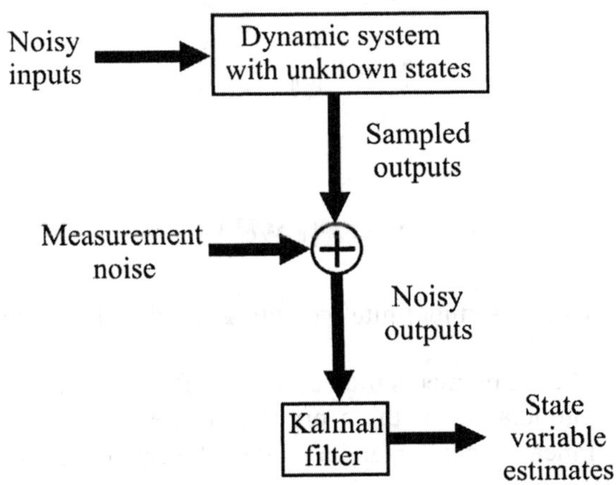

Figure 2.18 Kalman filter application.

extended Kalman filter (EKF) described later, the results are not optimum. The EKF is an ad hoc estimator because of the approximations employed.

2.5.4.1 Standard Kalman Filter

Denote the state vector with $\mathbf{s}_i \in \Re^N$ and the measurement vector with $\mathbf{z}_i \in \Re^P$. The Kalman filter assumes that the state vector follows the equation given by (2.115), and the measurements are given by (2.116).

The equations for the Kalman filter can be derived by a number of methods. Let $\hat{\mathbf{s}}_{i|-}$ denote the a priori state estimate at instant i given information about the process prior to instant i (that is, the measurements at instant i have not been factored in to update the state at instant i) and $\hat{\mathbf{s}}_i \in \Re^N$ denote the a posteriori state estimate at instant i, given measurement $\mathbf{z}_i$. Two estimate errors can then be defined as

$$\mathbf{e}_{i|-} \triangleq \mathbf{s}_i - \hat{\mathbf{s}}_{i|-} \tag{2.178}$$

and

$$\mathbf{e}_i \triangleq \mathbf{s}_i - \hat{\mathbf{s}}_i \tag{2.179}$$

which are the a priori and a posteriori estimate errors, respectively. The associated (weighted) error covariance matrices are

$$\mathbf{\Gamma}_{i|-} = \mathcal{E}\left\{\mathbf{e}_{i|-}\mathbf{W}\mathbf{e}_{i|-}^{\mathrm{T}}\right\} \tag{2.180}$$

and

$$\mathbf{\Gamma}_i = \mathcal{E}\left\{\mathbf{e}_i\mathbf{W}\mathbf{e}_i^{\mathrm{T}}\right\} \tag{2.181}$$

where $\mathbf{W}$ is any positive semidefinite weighting matrix, chosen to optimize some criterion.

The Kalman filter implements the derivation of an a posteriori estimate of the state as a linear combination of the a priori estimate and a weighted difference between the actual measurement $\mathbf{z}_k$ and a predicted measurement $\mathbf{H}\hat{\mathbf{x}}_{i|-}$:

$$\hat{\mathbf{s}}_i = \hat{\mathbf{s}}_{i|-} + \mathbf{K}\left(\mathbf{z}_i - \mathbf{H}\hat{\mathbf{s}}_{i|-}\right) \tag{2.182}$$

The last term in (2.182), ($\mathbf{z}_i - \mathbf{H}\hat{\mathbf{s}}_{i|-}$), is called the *innovation* of the process. Matrix **K**, which is $N \times P$, is selected to minimize the a posteriori error covariance (2.181). **K** is given by

$$\mathbf{K}_i = \mathbf{\Gamma}_{i|-}\mathbf{H}^{\mathrm{T}}\left(\mathbf{H}\mathbf{\Gamma}_{i|-}\mathbf{H}^{\mathrm{T}} + \mathbf{R}\right)^{-1} \tag{2.183}$$

The Kalman filter is a variable gain filter, as indicated by (2.183). As the measurement noise covariance is reduced ($\mathbf{R} \to \mathbf{0}$),

$$\lim_{\mathbf{R}\to 0} \mathbf{K}_i = \mathbf{\Gamma}_{i|-}\mathbf{H}^{\mathrm{T}}(\mathbf{H}^{\mathrm{T}})^{-1}\mathbf{\Gamma}_{i|-}^{-1}\mathbf{H}^{-1} = \mathbf{H}^{-1} \tag{2.184}$$

and the gain is adjusted with weighting favoring the residual—that is, more credibility is given to the measurements. On the other hand, as the a priori estimate error covariance $\mathbf{\Gamma}_{i|-}$ decreases,

$$\lim_{\mathbf{\Gamma}_{i|-}\to 0} \mathbf{K}_i = \mathbf{\Gamma}_{i|-}\mathbf{H}^{\mathrm{T}}\left(\mathbf{H}\mathbf{\Gamma}_{i|-}\mathbf{H}^{\mathrm{T}} + \mathbf{R}\right)^{-1} = \mathbf{0} \tag{2.185}$$

and the residual is weighted less heavily.

The *standard Kalman filter* is illustrated in Figure 2.19. It is defined by the evolution equations as follows (the numbers correspond to the numbers in the blocks in Figure 2.19):

- State prediction

$$(1) \qquad \hat{\mathbf{s}}_{i|-} = \mathbf{\Phi}\,\hat{\mathbf{s}}_{i-1} + \mathbf{B}\mathbf{u}_{k-1} \tag{2.186}$$

- Prediction of covariance matrix of states

$$(1) \qquad \mathbf{\Gamma}_{i|-} = \mathbf{\Phi}\,\mathbf{\Gamma}_{i-1}\mathbf{\Phi}^{\mathrm{T}} + \mathbf{Q} \tag{2.187}$$

- Kalman gain matrix computation

$$(2) \qquad \mathbf{K}_i = \mathbf{\Gamma}_{i|-}\mathbf{H}^{\mathrm{T}}\left(\mathbf{H}\,\mathbf{\Gamma}_{i|-}\mathbf{H}^{\mathrm{T}} + \mathbf{R}\right)^{-1} \tag{2.188}$$

- Update state estimation

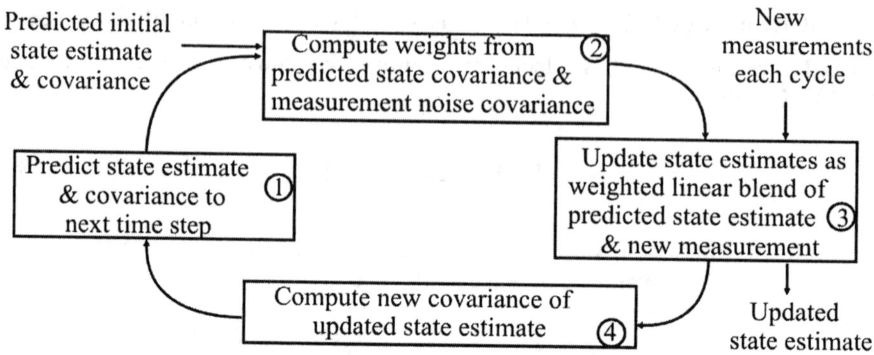

Figure 2.19 The standard Kalman filter algorithm.

$$\text{(3)} \qquad \hat{\mathbf{s}}_i = \hat{\mathbf{s}}_{i|-} + \mathbf{K}_i\left(\mathbf{z}_i - \mathbf{H}_i\hat{\mathbf{s}}_{i|-}\right) \tag{2.189}$$

- Update covariance matrix of states

$$\text{(4)} \qquad \mathbf{\Gamma}_i = (\mathbf{I} - \mathbf{K}_i\mathbf{H})\mathbf{\Gamma}_{i|-} \tag{2.190}$$

- Initialization

$$\begin{aligned} \mathbf{s}_{0|-} &= \mathcal{E}\{\mathbf{s}_0\} \\ \mathbf{\Gamma}_{0|-} &= \text{cov}\{\mathbf{s}_0\} \end{aligned} \tag{2.191}$$

Thus, the equations for the Kalman filter fall into two groups: time update equations and measurement update equations. The time update equations, given by (2.186) and (2.187), are responsible for projecting forward in time the current state and error covariance estimates to obtain the estimates for the next a priori time step. The measurement update equations, given by (2.189) and (2.190), are responsible for the feedback, that is, for incorporating a new measurement into the a priori estimate to obtain an improved a posteriori estimate.

Noise processes characterized by $\mathbf{Q}_i$ and $\mathbf{R}_i$ are measured or assumed to be of some value. They are often specified via an assumed value of SNR, and the SNR is varied to ascertain system performance trade-offs. As mentioned, frequently the noise processes are assumed to be uncorrelated so that

$$\mathbf{Q}_i = \text{diag}[\sigma_{n_1}^2 \quad \sigma_{n_2}^2 \quad \cdots \quad \sigma_{n_N}^2] \tag{2.192}$$

and

$$\mathbf{R}_i = \text{diag}[\sigma_{\eta_1}^2 \quad \sigma_{\eta_2}^2 \quad \cdots \quad \sigma_{\eta_P}^2] \tag{2.193}$$

Example: Suppose it is necessary to ascertain the diameter of a steel rod that is nominally 1 mm. Measurements of the diameter are sequentially made and processed in a Kalman filter. In this case, there is only one state variable: the diameter. Therefore $\mathbf{\Phi} = 1$, $\mathbf{B} = 0$, and

$$s_{i+1} = s_i \tag{2.194}$$

There is no process noise here, but suppose the manufacturer of the rod specifies that the diameter is 1 mm with a variance of 1% (0.01mm^2). Therefore, initially the best estimate of the diameter is $s_{0|-} = 1$ mm with an uncertainty variance of $\Gamma_{0|-} = 0.01$.

Repeated micrometer measurements

$$z_i = s_i + \eta_i \tag{2.195}$$

indicate the diameter to within the measurement noise η_i. Thus, $\mathbf{H} = 1$. It is assumed (reasonably) that the noise is uncorrelated from one measurement to the next. The micrometer manufacturer specifies that the micrometer is accurate to within an uncertainty variance of 0.1% (0.001 mm^2 in this case), and therefore $Q_0 = 0.001$.

Starting at $i = 0$, with the initial estimate of 1 mm and an uncertainty of 0.01, the weight for updating with the first measurement is

$$K_0 = \frac{\Gamma_{0|-}}{\Gamma_{0|-} + Q_0} = \frac{0.01}{0.01 + 0.001} = 0.91 \tag{2.196}$$

with an updated state estimate as

$$s_{0|0} = (1 - K_0)s_{0|-} + K_0 z_0 \tag{2.197}$$

where $s_{0|0}$ denotes the best estimate at time 0, based on the measurement at time 0. The associated variance of the updated estimate is

$$\Gamma_0 = (1 - K_0)\Gamma_{0|-} = (0.09)(0.01) = 9\times10^{-4} \tag{2.198}$$

According to (2.194), the state projects identically to time 1, so the state at time 1 and variance projection is

$$s_{1|0} = s_{0|0} \qquad \Gamma_{1|0} = \Gamma_{0|0} = 9\times10^{-4} \tag{2.199}$$

Repeating the cycle yields

$$K_1 = \frac{\Gamma_{1|0}}{\Gamma_{1|0} + Q_1} = \frac{9\times10^{-4}}{9\times10^{-4} + 10^{-3}} = 0.47 \tag{2.200}$$

and

$$\Gamma_{1|1} = (1 - K_1)\Gamma_{1|0} = (0.53)(9\times10^{-4}) = 4.8\times10^{-4} \tag{2.201}$$

Figure 2.20 illustrates a simulation of this process. The path of the covariance for this simulation is illustrated in Figure 2.21. The covariance gets to be quite small as the estimate approaches the true value of the diameter.

■

This discussion assumed that the Kalman filter is *linear, time invariant* (LTI). In general, nonlinear and time-variant versions of the Kalman filter are possible. In fact, it is these characteristics that make the Kalman filter as ubiquitous as it is. In particular, if $\mathbf{\Phi}_i\mathbf{s}_i$ is nonlinear or a linear relationship between $\mathbf{x}_i$ and $\mathbf{s}_i$ is not available, the *extended Kalman filter* (EKF) can be applied.

2.5.4.2 Extended Kalman Filter

The EKF approach is to apply the standard Kalman filter for the linear systems just discussed to nonlinear systems with AWGN by linearizing the nonlinearity via the Taylor series expansion and ignoring the terms of second order and higher. In other words, a linear Taylor approximation of the system function at the previous state estimate and that of the observation function at the corresponding predicted position are considered. Convergence to a reasonable estimate may not be obtained if the initial guess is poor or if the noise is so large that the linearization is inadequate to describe the system. Also, as mentioned, the EKF is an ad hoc

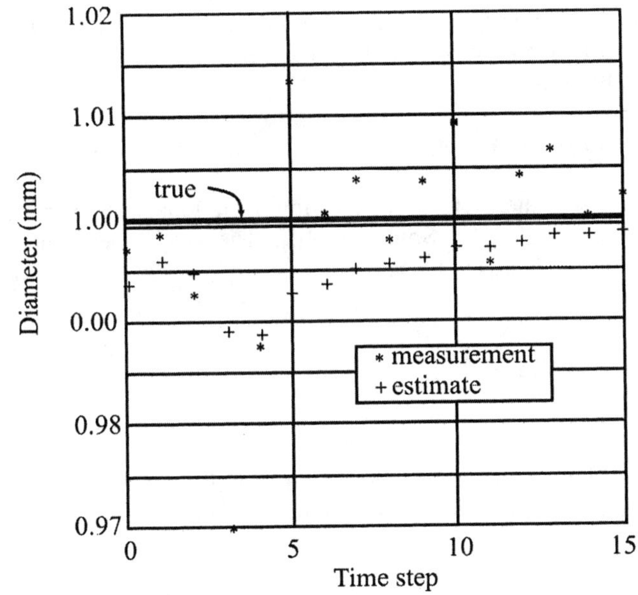

Figure 2.20 Diameter of the steel rod for the example.

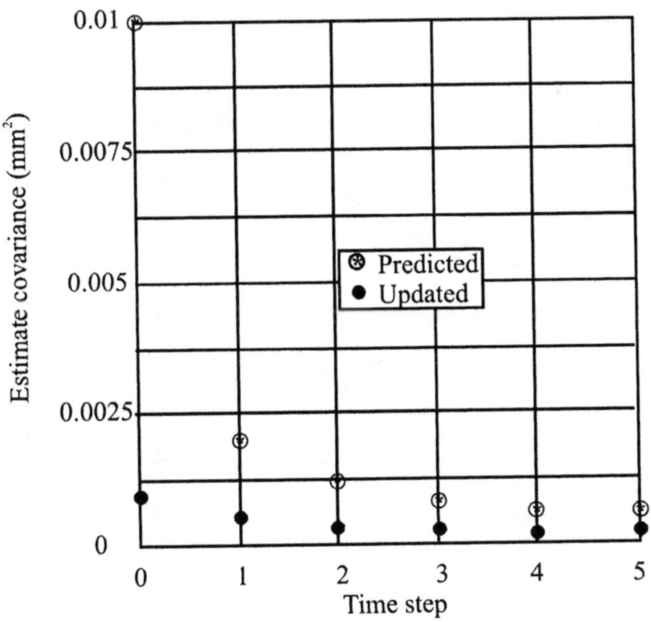

Figure 2.21 Variance of the rod diameter for the example.

estimator due to the approximations incorporated (in particular, the linearization).

It should be noted, however, that the EKF can have instability problems caused by the linearization approximation [13].

2.5.4.3 Position Fix Estimation with the Extended Kalman Filter

An analysis of using the EKF for the location of stationary emitters from a moving platform was conducted by Spingarn [14]. The two-dimensional geometry is illustrated in Figure 2.22. In this case,

$$\hat{\mathbf{x}}_k = \mathbf{f}(\hat{\mathbf{x}}_{k-1}, \mathbf{0}, \mathbf{0}) = \begin{bmatrix} \hat{x}_{\mathrm{T},k} \\ \hat{y}_{\mathrm{T},k} \end{bmatrix} = \begin{bmatrix} 1 & 0 \\ 0 & 1 \end{bmatrix} \begin{bmatrix} \hat{x}_{\mathrm{T},k-1} \\ \hat{y}_{\mathrm{T},k-1} \end{bmatrix} \tag{2.202}$$

The state transition matrix in this case is the identity matrix because the target is stationary. If the target were moving, then there would be dynamic terms involved.

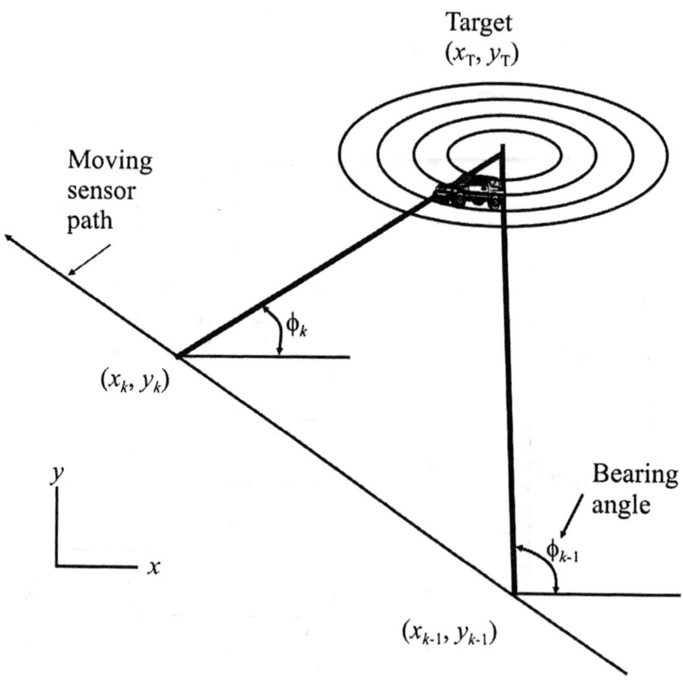

Figure 2.22 Geometry for EKF PF analysis.

The state transition in this case is linear. The observables are the bearing angles ϕ_i and the measurement model is given by

$$\phi_k = h(\hat{\mathbf{x}}_k) + \eta_k \tag{2.203}$$

where η_k represents the noise process, assumed here to be zero mean AWGN such that

$$\mathcal{E}\{\eta_j \eta_k\} = \sigma_\eta^2 \delta_{jk} \tag{2.204}$$

and

$$\mathbf{R} = \sigma_\eta^2 \tag{2.205}$$

The state estimation equation is

$$\hat{\mathbf{x}}_k = \hat{\mathbf{x}}_{k|-} + \mathbf{K}\left[\phi_k - h\left(\hat{\mathbf{x}}_{k|-}\right)\right] \tag{2.206}$$

where

$$h\left(\hat{\mathbf{x}}_{k|-}\right) = \hat{\phi}_k = \tan^{-1}\left(\frac{\hat{y}_{\mathrm{T},k|-} - y_k}{\hat{x}_{\mathrm{T},k|-} - x_k}\right) \tag{2.207}$$

The update for the covariance matrix of states is

$$\mathbf{\Gamma}_k = \left[\mathbf{I} - \mathbf{K}\mathbf{H}\left(\hat{\mathbf{x}}_{k|-}\right)\right]\mathbf{\Gamma}_{k|-} \tag{2.208}$$

where **H** is the Jacobian matrix corresponding to h:

$$\begin{aligned} \mathbf{H}\left(\hat{\mathbf{x}}_{k|-}\right) &= \left.\frac{\partial h(\mathbf{x})}{\partial \mathbf{x}}\right|_{\mathbf{x}=\mathbf{x}_{k|-}} \\ &= \begin{bmatrix} H_{11} & H_{12} \end{bmatrix} \end{aligned} \tag{2.209}$$

where

$$H_{11} = \left.\frac{\partial h}{\partial x_{\mathrm{T}}}\right|_{x_{\mathrm{T}}=\hat{x}_{\mathrm{T},k|-}} = \frac{-u}{1+u^2}\frac{1}{\hat{x}_{\mathrm{T},k|-} - x_k} \tag{2.210}$$

and

$$H_{12} = \left.\frac{\partial h}{\partial y_{\mathrm{T}}}\right|_{y_{\mathrm{T}}=\hat{y}_{\mathrm{T},k|-}} = \frac{1}{1+u^2}\frac{1}{\hat{x}_{\mathrm{T},k|-} - x_k} \tag{2.211}$$

with

$$u = \frac{\hat{y}_{\mathrm{T},k|-} - y_k}{\hat{x}_{\mathrm{T},k|-} - x_k} \tag{2.212}$$

The Kalman gain matrix is

$$\mathbf{K} = \mathbf{\Gamma}_{k|-}\mathbf{H}^{\mathrm{T}}\left(\hat{\mathbf{x}}_{k|-}\right)\left[\mathbf{H}\left(\hat{\mathbf{x}}_{k|-}\right)\mathbf{\Gamma}_{k|-}\mathbf{H}^{\mathrm{T}}\left(\hat{\mathbf{x}}_{k|-}\right) + \mathbf{R}_k\right]^{-1} \tag{2.213}$$

where

$$\mathbf{R}_k = \begin{bmatrix} \sigma_x^2 & 0 \\ 0 & \sigma_y^2 \end{bmatrix} \tag{2.214}$$

and the error covariance is

$$\mathbf{\Gamma}_{k+1} = \mathbf{J}\mathbf{\Gamma}_k\mathbf{J}^{\mathrm{T}} + \mathbf{Q} \tag{2.215}$$

where

$$\mathbf{Q} = \begin{bmatrix} q_{11} & 0 \\ 0 & q_{22} \end{bmatrix} \tag{2.216}$$

Q is a function of the noise. In this case, there is no noise in the model, so $q_{11} = 0$ and q_{22} is set to a small positive value.

In this development, the state estimate equation is linear but the measurement equation is a $\tan^{-1}$ function—highly nonlinear. Initial values of the state variables (location) are required, denoted by $\hat{\mathbf{x}}_{e0}$. Initial values for the error covariance are also required:

$$\mathbf{\Gamma}_{0|0} = \begin{bmatrix} \sigma_x^2 & 0 \\ 0 & \sigma_y^2 \end{bmatrix} \tag{2.217}$$

Example: Example results of a Monte Carlo simulation of the EKF model are shown in Figure 2.23. For this figure, $\sigma_n^2 = (0.0175)^2 \text{ rad}^2 = (1^\circ)^2$. The target was located at (141, 141) miles and the initial position of the aircraft was (x, y) = (130, 11) miles. The velocity of the sensor was v = 0.1 miles/sec. Measurements were made every 300 sec and the closest the sensor got to the target was 100 miles. The bearing error was caused by AWGN with $\sigma = 1^\circ$. The initial estimate of target location to initialize the iterations was $\hat{x}_{T,0} = \hat{y}_{T,0} = 135 \text{ miles}$, q_{11} = 0 (because there is no process noise included), q_{22} = 0.0252 rad^2, and

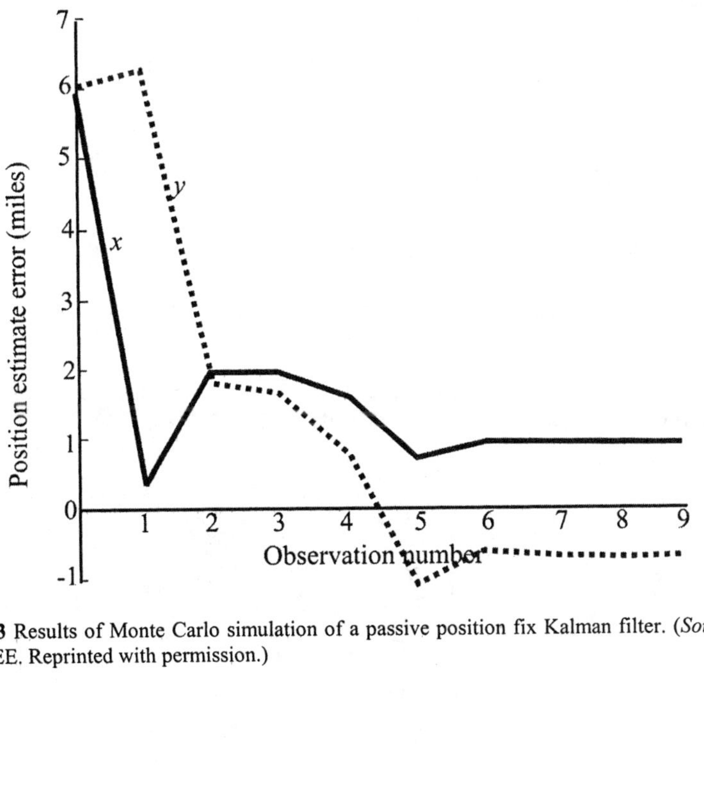

Figure 2.23 Results of Monte Carlo simulation of a passive position fix Kalman filter. (*Source:* [14], © 1987 IEEE. Reprinted with permission.)

$\Gamma_{0|-} = \text{diag}(5^2)\,\text{miles}^2$. The x-error converges to about 1 mile, while y-error converges to about –0.7 mile after nine observations.

■

2.5.4.4 Total Least-Squared with a Kalman Filter

If the bearing errors contain a bias Φ, as expressed in (2.72), the TLS solution is not necessarily the best statistical estimate. An alternate technique using TLS and a Kalman filter is discussed in this section [11]. This Kalman filter approach deviates somewhat from the earlier development in Section 2.5.3.1, which serves to illustrate the considerable flexibility available with the Kalman filter.

The bearing errors have a constant part given by Φ and a random part given by $\Delta\phi_i^*$:

$$\phi_i = \Phi + \Delta\phi_i^* \tag{2.218}$$

First the TLS estimate is found, and that estimate is used in (2.68). The state equations of the filter are given by

$$\delta\phi_{i+1} = \delta\phi_i \tag{2.219}$$

$$\Phi_{i+1} = \Phi_i \tag{2.220}$$

while the observation equation is

$$\tilde{\phi}_i - \hat{\phi}_i = \Delta\hat{\phi}_i = \delta\phi_i + \Phi_i + \Delta\phi_i^* \tag{2.221}$$

where $\delta\phi_i$ is the bearing error at interval i due to the error in the estimation of (ψ_T, λ_T).

The Kalman filter algorithm is found as follows. Suppose that M observations are available.

For $i = 1, 2, ..., M$

$$\begin{bmatrix} \hat{\phi}_i \\ \delta\hat{\phi}_i \end{bmatrix} = \begin{bmatrix} \hat{\phi}_{i-1} \\ \delta\hat{\phi}_{i-1} \end{bmatrix} + \mathbf{K}_i \left(\Delta\hat{\phi}_i - \mathbf{C} \begin{bmatrix} \hat{\Phi}_{i-1} \\ \delta\hat{\Phi}_{i-1} \end{bmatrix} \right)$$

$$\mathbf{K}_i = \mathbf{\Gamma}_{i-1} \frac{\mathbf{C}^\text{T}}{R_{2_i} + \mathbf{C}\mathbf{\Gamma}_{i-1}\mathbf{C}^\text{T}}$$

$$\mathbf{\Gamma}_i = \mathbf{\Gamma}_{i-1} - \mathbf{\Gamma}_{i-1}\mathbf{C}^{\mathrm{T}}\mathbf{C}\frac{\mathbf{\Gamma}_{i-1}}{R_{2_i} + \mathbf{C}\mathbf{\Gamma}_{i-1}\mathbf{C}^{\mathrm{T}}}$$

$\mathbf{K}_i$ is the Kalman gain vector
R_{2_i} is the covariance of $\Delta\phi_i^*$
$\mathbf{\Gamma}_i$ is the covariance matrix
$\mathbf{C} = [1\ 1]$

The set of observations $\tilde{\tilde{\phi}}_i = \tilde{\phi}_i - \hat{\phi}_i$ for all i are then used with the TLS algorithm above to obtain a new estimate $(\tilde{\tilde{\psi}}_{\mathrm{T}}, \tilde{\tilde{\lambda}}_{\mathrm{T}})$. These values are then used in (2.105) to generate a new estimate of θ_i. An updated estimate $\tilde{\tilde{\Phi}}_n$ is then obtained with the Kalman filter. This procedure is repeated until the bias disappears.

Example: In addition to conducting a simulation experiment for TLS PF computations, Rao and Reddy [11] also included a simulation experiment using TLS with the Kalman filter. The geometry was as indicated in Figure 2.13, and 122 bearings were taken on the baseline length of 242 km. The results are indicated in Table 2.2. This indicates that TLS alone is not necessarily optimal when there is a systematic bias in the bearings. Using TLS with the Kalman filter produced significantly better results in that case.

■

2.5.4.5 Unscented Kalman Filter

The EKF linearizes the system nonlinearity by expanding it in a Taylor series and retaining only the linear terms. When the nonlinearity is highly nonlinear, retaining only the linear terms can lead to significant modeling errors and divergent behavior. An extension to the EKF was proposed by Julier and Uhlman [15] called the *unscented Kalman Filter* (UKF). This method improves on the performance of the EKF by, inter alia, retaining up through the third-order nonlinear terms in the Taylor series. This, coupled with retaining only specially

Table 2.2 Results of TLS Estimation with a Kalman Filter

Case	Random error	Systematic error	Error (km)
1	$\sim\mathcal{N}(0, 1)$	0°	0.524
2	$\sim\mathcal{N}(0, 1)$	–3°	0.524
3	$\sim\mathcal{N}(0, 1)$	6°	0.524

Source: [11].

selected sample points to propagate from one state to the next, can lead to much better models for nonlinear systems.

2.6 The Discrete Probability Density Method

All of the position fixing algorithms discussed here are subject to erroneous results due to errors in the data used to compute the fixes. The effects of errors vary. Presented in this section is a technique that mitigates some of the erroneous results when the input data for the fix computations are LOBs [16].

The AOI is segmented into small squares or rectangles with a grid such as the one shown in Figure 2.24. This grid can be as granular as desired/required; however, the smaller the segmentation the higher the computational burden and

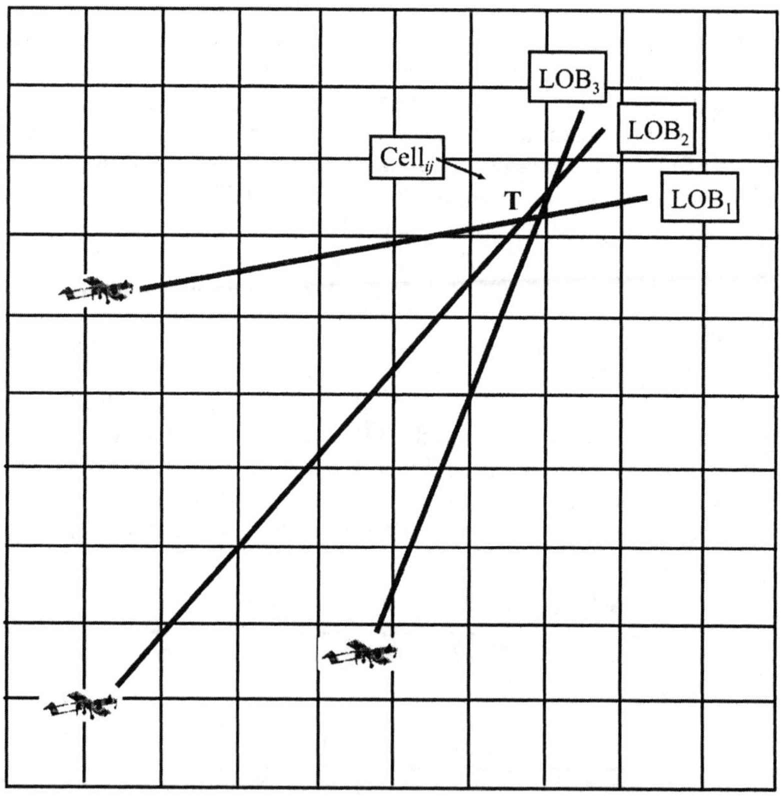

Figure 2.24 DPD grid with three LOBs.

the higher the resolution (but not necessarily the accuracy). A 100 × 100 grid requires 100 times the amount of computations of a 10 × 10 grid, for example.

The position-fixing algorithm computes the probability that an emitter lies within each of the grid segments based on the LOBs provided. The LOBs are assumed to have a von Mises pdf (also called the Tikhonov pdf [17]) that is defined over the angular interval (–π, π). The Tikhonov pdf is given by [16]

$$p(\phi)=\begin{cases}\dfrac{\exp[\nu\cos(\phi-\mu)]}{2\pi I_0(\nu)}, & -\pi<\phi\leq\pi\\ \text{undefined}, & \text{elsewhere}\end{cases} \tag{2.222}$$

where ν is the *concentration* and indicates the dispersion of ϕ around μ and is similar to the inverse of the variance of the Gaussian density. μ is the mean of the distribution and, for the purposes here, is the measured LOB, which ranges over (–π, π] radians. This function is plotted in Figure 2.25 for some values of ν. As ν gets larger, the Tikhonov pdf gets larger at the mean value (zero in this case) and narrower, and the variance is smaller.

The distribution function corresponding to (2.222) is given by

$$P(\phi)=\frac{1}{2\pi I_0(\nu)}\left\{\phi I_0(\nu)+2\sum_{i=1}^{\infty}\frac{I_i(\nu)\sin[i(\phi-\mu)]}{i}\right\} \tag{2.223}$$

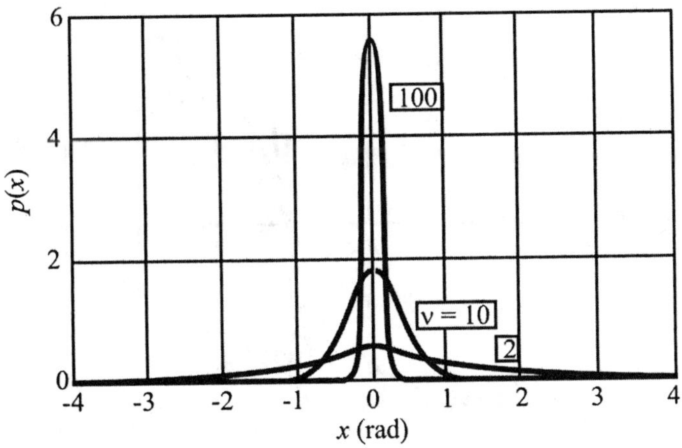

Figure 2.25 von Mises pdf.

which is plotted in Figure 2.26 for the same values. As expected and predicted from Figure 2.25, the larger ν is, the more rapid is the transition from low to high probability around the mean value.

The pdf expands as the distance from the sensor increases. Close to the sensor, the pdf is narrow, with probability values close to the mean as illustrated in Figure 2.27. Farther from the sensor, the pdf expands, encompassing more grid squares.

The errors in the LOBs from different platforms are independent random variables, so the overall pdf obtained by combining N individual distributions is given by

$$F(x,y) = C\prod_{i=1}^{N} \frac{\exp[\nu_i \cos(\phi_i - \mu_i)]}{2\pi I_0(\nu_i)} \tag{2.224}$$

The constant C is present to normalize the total volume under the total grid surface and is present to avoid numerical stability problems. It is calculated after all the probability terms have been determined. After normalization, the probability value for each cell is given by $F(x, y)$.

Based on this pdf, the probability that the target lies within a particular segment is computed by first computing the probabilities at each of the corners of the grid segment. The overall probability of being within the segment is then the average of these probabilities, to wit,

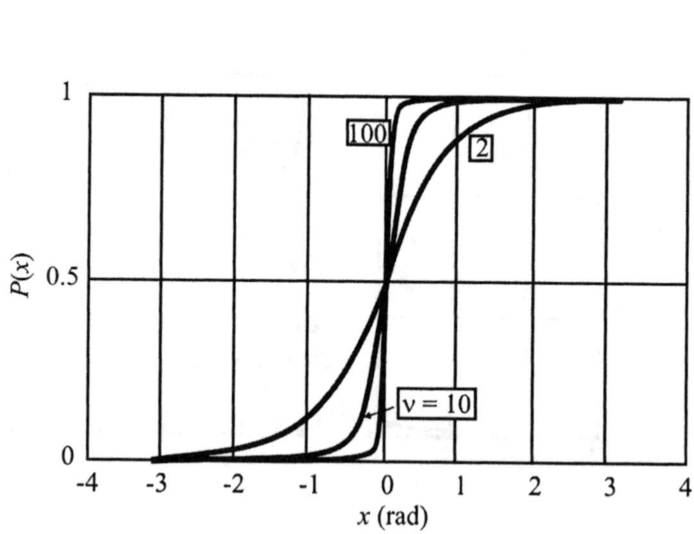

Figure 2.26 von Mises distribution function.

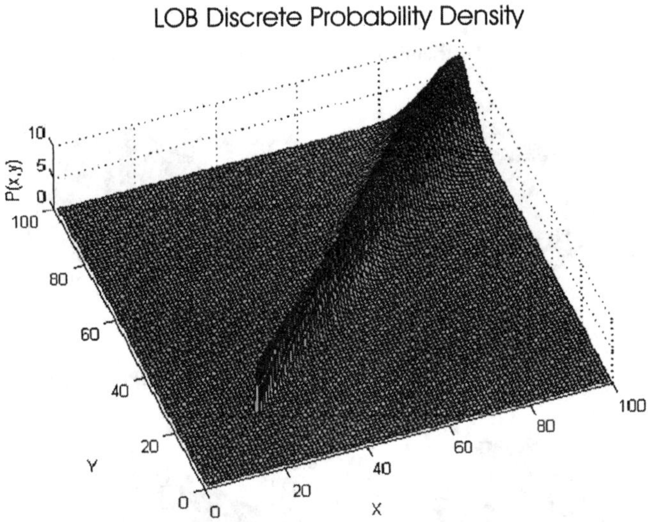

Figure 2.27 DPD LOB distribution. (*Source:* [16], © 2003 British Crown. Reprinted with permission.)

$$\Pr(\text{within cell}_{i,j}) = \frac{1}{4}[F(x_i, y_j) + F(x_{i+1}, y_j) + F(x_i, y_{j+1}) + F(x_{i+1}, y_{j+1})] \quad (2.225)$$

A wild bearing tends to have less effect on this method of fix computation. This is because the overall fix pdf is the product of the *N* individual pdfs. Even if the wild bearing has a small associated variance, its contribution to the overall pdf will be small because the wild pdf is largest where the tails of the good pdfs are located. Assuming the remainder of the pdfs are of good quality, their product will be larger than the product of the wild pdf main lobe and the good pdf tails, even though the former pdf is multiplied by the wild pdf tail.

Example: Figure 2.28 illustrates the technique for three PF sensors located according to the data in Table 2.3. The product of the pdfs is significant only over a small segment of the grid, as shown in Figure 2.28(d). In this case, the *root mean- squared* (RMS) error is 3°, corresponding to

$$\nu = \frac{1}{\left(3\frac{\pi}{180}\right)^2} = 364.8 \quad (2.226)$$

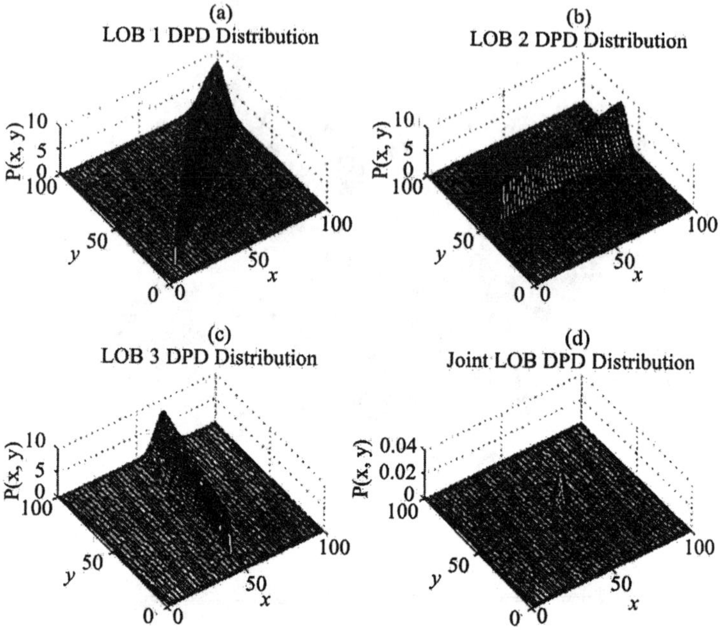

Figure 2.28 Example of the DPD PF algorithm with three PF stations. (*Source*: [16], © 2003 British Crown. Reprinted with permission.)

Table 2.3 Bearings for the DPD Example

PF System	X-coordinate	Y-coordinate	Bearing from y-axis (degrees)
1	10	10	45
2	10	50	80
3	50	10	10

The volume under the curve in Figure 2.28(d) between $55 < x < 65$ and $55 < y < 65$ is approximately 0.92 and therefore approximately equal to the 90% error ellipse. This ellipse has a major axis of 14 cells and a minor axis of 8 cells, with a major axis orientation of 45°.

■

2.7 Generalized Bearings

A different approach to iterating toward a PF solution was presented by Paradowski [18]. Given the ith azimuth AOA ϕ_i and the ith elevation AOA φ_i, the *generalized bearing* α_i is determined as indicated in Figure 2.29. The generalized bearing is in the plane defined by triangle TS_iA and is given by

$$f_i(\mathbf{x}_i) = \alpha_i = \tan^{-1} \frac{\sqrt{(x_\text{T} - x_i)^2 + (z_\text{T} - z_i)^2}}{y_\text{T} - y_i} \tag{2.227}$$

The measurements are expressed as

$$u_i = f_i(\mathbf{x}_i) + n_i, \qquad i = 1, 2, \ldots, M \tag{2.228}$$

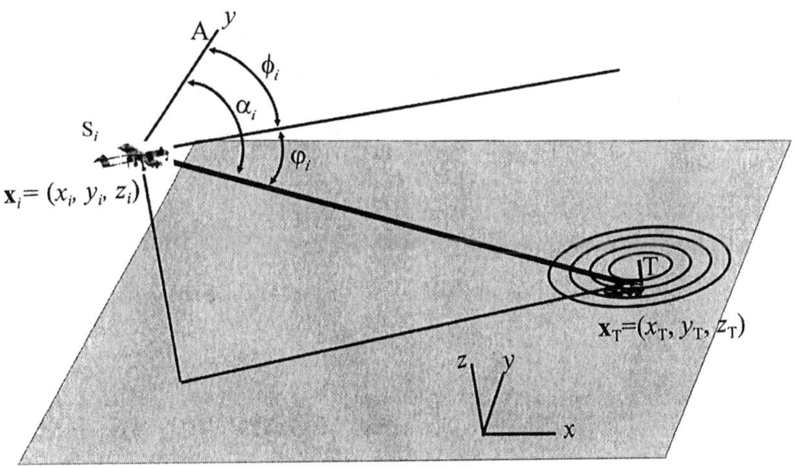

Figure 2.29 Generalized bearing, denoted by α_i.

or

$$\mathbf{u} = \mathbf{f}(\mathbf{x}) + \mathbf{n} \tag{2.229}$$

where, in the absence of measurement errors $u_i = \mathrm{f}_i(\mathbf{x}i) = \alpha_i$, and $\mathrm{f}_i(\mathbf{x}_i)$ is the function in which the PF and LOB information is contained. The measurement noise, n_i, is characterized by

$$\mathcal{E}\{n_i\} = 0 \qquad \mathcal{E}\{n_i n_j\} = \sigma_{ij} \tag{2.230}$$

$$\sigma_{ij} = \begin{cases} \sigma_{ij}^2 & i = j \\ \rho_{ij}\sigma_i\sigma_j & i \neq j \end{cases} \tag{2.231}$$

σ_i and σ_j are standard deviations for the ith and jth measurements and ρ_{ij} is the correlation coefficient between these measurements.

The gradient vector at the mth step in the iteration toward the solution is given by

$$\mathbf{g}_{m_i} = \begin{bmatrix} \dfrac{\hat{x}_m - x_i}{r_{m_i}^2 s_{m_i}} & -\dfrac{\hat{y}_m - y_i}{r_{m_i}} s_{m_i} & \dfrac{\hat{z}_m - z_i}{r_{m_i}^2 s_{m_i}} \end{bmatrix}^{\mathrm{T}} \tag{2.232}$$

where

$$r_{m_i}^2 = (\hat{x}_m - x_i)^2 + (\hat{y}_m - y_i)^2 + (\hat{z}_m - z_i)^2 \tag{2.233}$$

and

$$s_{m_i} = \sqrt{\left(\frac{r_{m_i}}{\hat{y} - y_i}\right)^2 - 1} \tag{2.234}$$

The gradient matrix consists of rows that are the gradient vectors

$$\mathbf{G}_m = \begin{bmatrix} \mathbf{g}_{m1} & \mathbf{g}_{m2} & \cdots & \mathbf{g}_{mM} \end{bmatrix}^{\mathrm{T}} \tag{2.235}$$

At the mth iteration, the covariance matrix of generalized bearing errors is a diagonal matrix expressed as

$$\mathbf{R}_m = \text{diag}[\sigma_{m_1}^2, \sigma_{m_2}^2, \ldots, \sigma_{m_M}^2] \tag{2.236}$$

where $\sigma_{m_i}^2$ is the variance of the α_i error at the mth iteration. These can be expressed in terms of the variances of the measurements as

$$\sigma_{m_i}^2 = a_{m_i}\sigma_{\phi_i}^2 + b_{m_i}\sigma_{\varphi_i}^2 + c_{m_i}\rho_i\sigma_{\phi_i}\sigma_{\varphi_i} \tag{2.237}$$

where

$$a_{m_i} = \left(\frac{1}{s_{m_i}}\frac{\hat{x}_m - x_i}{\hat{y}_m - y_i}\right)^2$$

$$b_{m_i} = \left\{s_{m_i}\left[\left(\frac{r_{m_i}}{\hat{z}_m - z_i}\right) - 1\right]^{1/2}\right\}^{-2}$$

$$c_{m_i} = \sqrt{a_{m_i} b_{m_i}}$$

Given that the noise is AWGN, then the iteration equation is given by

$$\hat{\mathbf{x}}_{m+1} = (\mathbf{G}_m^{\mathrm{T}}\mathbf{R}_m^{-1}\mathbf{G}_m)^{-1}\mathbf{G}_m^{\mathrm{T}}\mathbf{R}_m^{-1}\mathbf{w}_m \tag{2.238}$$

where

$$\mathbf{w}_m = \mathbf{u} + \mathbf{G}_m\hat{\mathbf{x}}_m - \mathbf{f}(\hat{\mathbf{x}}_m) \tag{2.239}$$

and where $\mathbf{f}(\mathbf{x}_m)$ is $\mathbf{f}(\mathbf{x})$ evaluated at $\mathbf{x} = \mathbf{x}_m$.

Of course, being an iterative procedure, and an approximate procedure, a good initial guess is required or the solution may settle to a local minimum. Also, convergence cannot be guaranteed.

Example: Consider the geometry shown in Figure 2.30. The unattended aerial vehicle (UAV) sensor is flying along the x-axis at an altitude of 3,000m and the targets are somewhere in the 20 × 20-km region. The y-coordinate of the sensor is therefore unchanging and set to zero. The sensor makes an LOB measurement every 1 km along its straight route.

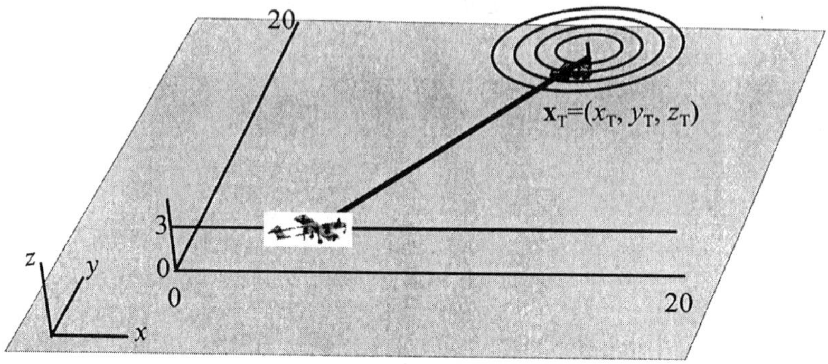

Figure 2.30 Example scenario for the generalized bearing algorithm.

The results of this algorithm for two target locations are shown in Figure 2.31. The results appear to be converging to the correct location, although there were many cases that did not converge.

The parameters used in the example cause the convergence to be particularly sensitive to $\hat{z}_0$. This is likely due to the low altitude at which the sensor was flown compared to the distance to the targets; the slant angle was fairly low.

■

A MathCad® program listing is contained in Appendix B of this chapter that implements one pass through the iterative procedure for the generalized bearing algorithm described.

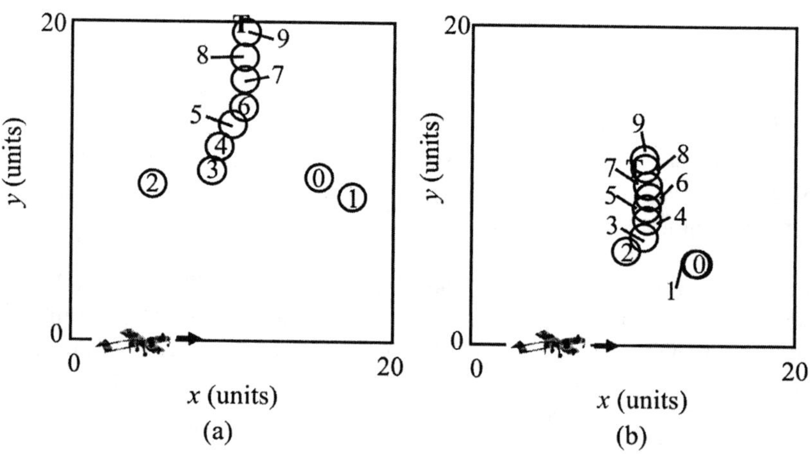

Figure 2.31 Results of the generalized bearing algorithm example. The target in (a) is located at $\mathbf{x}_T$ = (10, 20, 0.5), while the target in (b) is at (10, 10, 0.5). Variations in $\hat{z}$ are not shown.

2.8 Maximum Likelihood PF Algorithm

The analysis of interferometric LOB algorithms using a maximum likelihood approach is presented in this section. The development follows that in [19] closely.

Consider the geometry shown in Figure 2.32, which contains an airborne sensor flying in the $+y$ direction with velocity v. It contains two receivers as noted. Distance y_k is at the center of the two receivers at instant k. Measurements are taken over the observation interval, which has a center located at y_c. The target is located at coordinates (x_T, y_T) (the z-dimension is ignored for now). Initially it is assumed that $y_T = 0$. $M = 2P + 1$ measurements are collected over interval L.

The sensor measures the phase difference of the signals received by the two receivers. This difference is related to the ranges r_1 and r_2 by

$$\Delta\phi_i = 2\pi\frac{\Delta r}{\lambda} \tag{2.240}$$

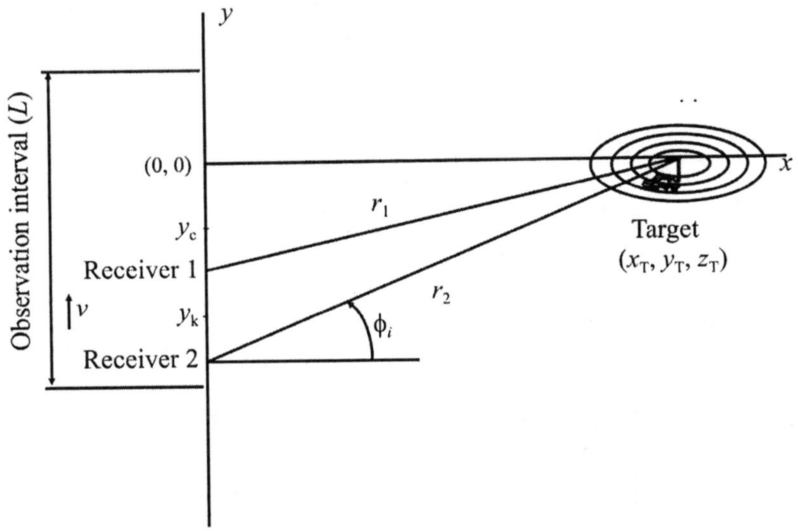

Figure 2.32 Geometry for the maximum likelihood algorithm. (*Source:* [19], © 1989 IEEE. Reprinted with permission.)

When the baseline between the two receivers is small compared with the range to the target and the bearing angle is small, the two rays that delineate r_1 and r_2 in Figure 2.32 are approximately parallel, as shown in Figure 2.33. Then

$$\sin\phi = \frac{\Delta r}{b}$$

but for small ϕ,

$$\sin\phi \approx \phi$$

so

$$\phi \approx \frac{\Delta r}{b}$$

so the error in the range difference, as indicated by the standard deviation $\sigma_{\Delta r}$, is related to the error in the angle measurement as indicated by the standard deviation σ_ϕ as

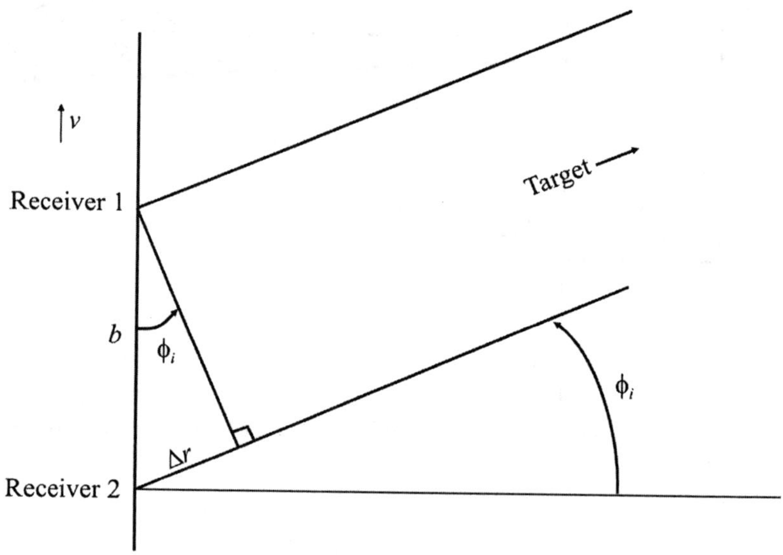

Figure 2.33 Details of the geometry shown in Figure 2.17. (*Source:* [19], © 1989 IEEE. Reprinted with permission.)

$$\sigma_\phi \approx \frac{\sigma_{\Delta r}}{b}, \qquad \phi << 1 \text{ rad} \tag{2.241}$$

The noise is assumed to be Gaussian with zero mean and variance σ_n^2. The estimator used in this PF system is a least-squared estimator that implements an iterative approach.

2.8.1 Maximum Likelihood Estimation Triangulation Algorithm

The unknowns in the geometry shown in Figure 2.32 are y_c, the center of the observation interval, and x_T, the location of the target (since it is assumed for now that $y_\text{T} = 0$). Therefore, the vector of unknowns is

$$\mathbf{x} = \begin{bmatrix} x_\text{T} & y_\text{c} \end{bmatrix}^\text{T} \tag{2.242}$$

For the interferometer, from Figure 2.32, the noiseless measurements are

$$\tilde{h}_k(x_\text{T}, y_\text{c}) = \phi_k = \tan^{-1}\left(-\frac{y_k}{x_\text{T}}\right) \tag{2.243}$$

where

$$y_k = y_\text{c} + k\Delta L$$

The $2P + 1$ noisy measurements, denoted by z_k, corrupted by samples of the Gaussian noise process denoted by n_k, are given by

$$z_k(x_\text{T}, y_\text{c}) = \tilde{z}_\text{k}(x_\text{T}, y_\text{c}) + n_k, \qquad k = 0, \pm 1, \pm 2, \ldots, \pm P \tag{2.244}$$

Let

$$\partial_{kx} = \frac{\partial \tilde{z}_k(x_\text{T}, y_\text{c})}{\partial x_\text{T}} \tag{2.245}$$

$$\partial_{ky} = \frac{\partial \tilde{z}_k(x_\text{T}, y_\text{c})}{\partial y_\text{c}} \tag{2.246}$$

Define the matrix of partial derivatives as

$$\mathbf{G} = \begin{bmatrix} \partial_{-Px} & \cdots & \partial_{kx} & \cdots & \partial_{Px} \\ \partial_{-Py} & \cdots & \partial_{ky} & \cdots & \partial_{Py} \end{bmatrix}^{\mathrm{T}} \tag{2.247}$$

and

$$\mathbf{z} = [z_{-P} \quad \cdots \quad z_k \quad \cdots \quad z_P]^{\mathrm{T}} \tag{2.248}$$

$$\tilde{\mathbf{z}} = [\tilde{z}_{-P} \quad \cdots \quad \tilde{z}_k \quad \cdots \quad \tilde{z}_P]^{\mathrm{T}} \tag{2.249}$$

$$\mathbf{n} = [n_{-P} \quad \cdots \quad n_k \quad \cdots \quad n_P]^{\mathrm{T}} \tag{2.250}$$

$$\mathbf{w} = [\Delta x_{\mathrm{T}} \quad \Delta y_{\mathrm{c}}]^{\mathrm{T}} \tag{2.251}$$

If the noise is i.i.d. with covariance matrix

$$\mathbf{Cov(n)} = \sigma_n^2 \mathbf{I} \tag{2.252}$$

then the correction vector for the next iteration is given by

$$\mathbf{w} = (\mathbf{G}^{\mathrm{T}}\mathbf{G})^{-1}(\mathbf{z} - \tilde{\mathbf{z}}) \tag{2.253}$$

and the estimate for the next iteration is

$$\mathbf{x}_{i+1} = \mathbf{x}_i + \mathbf{w} \tag{2.254}$$

The iterating ceases when **w** vanishes.

The final covariance matrix for the variables gives the PF error. It is given by

$$\mathbf{Cov(x)} = \sigma_n^2 (\mathbf{G}^{\mathrm{T}}\mathbf{G})^{-1} \tag{2.255}$$

From (2.243), (2.245), and (2.246),

$$\partial_{kx} = \frac{y_{\mathrm{c}} + k\Delta L}{x_{\mathrm{T}}^2 + (y_{\mathrm{c}} + k\Delta L)^2} \tag{2.256}$$

and

$$\partial_{ky} = \frac{-x_{\mathrm{T}}}{x_{\mathrm{T}}^2 + (y_{\mathrm{c}} + k\Delta L)^2} \tag{2.257}$$

so that

$$\mathbf{G}^{\mathrm{T}}\mathbf{G} = \begin{bmatrix} \sum_{k=-P}^{P} (\partial_{kx})^2 & \sum_{k=-P}^{P} \partial_{kx}\partial_{ky} \\ \sum_{k=-P}^{P} \partial_{kx}\partial_{ky} & \sum_{k=-P}^{P} (\partial_{ky})^2 \end{bmatrix} \tag{2.258}$$

When $y_{\mathrm{c}} = 0$, the off diagonal terms in (2.258) are zero, so

$$\operatorname{var}^{-1} x_{\mathrm{T}} = \frac{1}{\sigma_{\phi}^2} \sum_{k=-P}^{P} \frac{(k\Delta L)^2}{[x_{\mathrm{T}}^2 + (k\Delta L)^2]^2} \tag{2.259}$$

and

$$\operatorname{var}^{-1} y_{\mathrm{c}} = \frac{1}{\sigma_{\phi}^2} \sum_{k=-P}^{P} \frac{x_{\mathrm{T}}^2}{[x_{\mathrm{T}}^2 + (k\Delta L)^2]^2} \tag{2.260}$$

Again, assuming that the observations are dense, so that $\Delta L \approx dL$, multiplying by dL inside the sum, dividing by ΔL outside the sum, and replacing the sum with an integral with

$$k\Delta L = y, \qquad P\Delta L = \frac{L}{2} \tag{2.261}$$

then

$$\operatorname{var}^{-1} x_{\mathrm{T}} = \frac{1}{\sigma_{\phi}^2 \Delta L} \int_{-L/2}^{L/2} \frac{y^2}{[x_{\mathrm{T}}^2 + y^2]^2} dy \tag{2.262}$$

and

$$\text{var}^{-1}\, y_{\text{c}} = \frac{1}{\sigma_\phi^2 \Delta L} \int_{-L/2}^{L/2} \frac{x_{\text{T}}^2}{[x_{\text{T}}^2 + y^2]^2} dy \tag{2.263}$$

Assuming $x_{\text{T}} >> L$ and the fact that

$$M = 2P + 1 \approx 2P = \frac{L}{\Delta L} \tag{2.264}$$

then (2.262) and (2.263) become

$$\text{var}^{-1} x_{\text{T}} \approx \frac{ML^2}{12\sigma_\phi^2 x_{\text{T}}^4} \tag{2.265}$$

and

$$\text{var}^{-1} y_{\text{c}} \approx \frac{M}{\sigma_\phi^2 x_{\text{T}}^2} \tag{2.266}$$

so that

$$\sigma_x \approx \sqrt{\frac{12}{M}} \frac{\sigma_\phi x_{\text{T}}^2}{L} \tag{2.267}$$

and

$$\sigma_y \approx \frac{\sigma_\phi x_{\text{T}}}{\sqrt{M}} \tag{2.268}$$

When $y_{\text{c}} \neq 0$, (2.267) remains the same while (2.268) changes to

$$\sigma_y \approx \sqrt{\frac{1}{M}} \sigma_\phi x_{\text{T}} \sqrt{1 + 12 \frac{y_c^2}{L^2}} \tag{2.269}$$

Example: The performance of the LOB algorithm is illustrated in Figure 2.34, where x_0 = 10 km, L = 1.6 km, M = 40, $\sigma_{\Delta_r} = 0.0141$m, and the initial guess of b = 50m. The down-range standard deviation remained constant as dictated by (2.267), while the cross-range standard deviation varied as y_c was varied, as expected from (2.269).

■

2.8.2 Maximum Likelihood Estimation Algorithm Comparison

Gavish and Weiss compared the performance of two MLE PF algorithms [20]. Assuming the noise is Gaussian with zero mean, the *maximum likelihood*

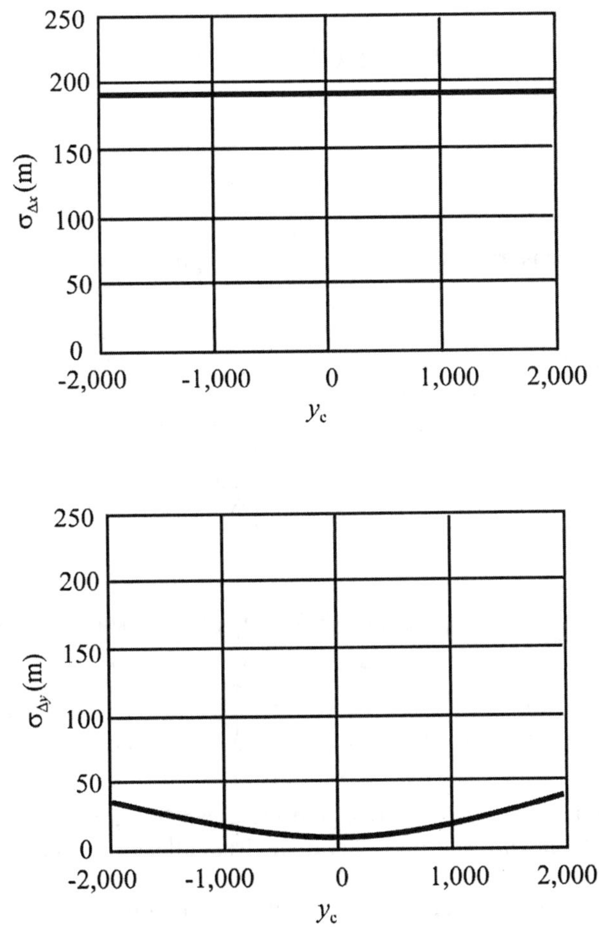

Figure 2.34 Interferometer results. (*Source:* [19], © 1989 IEEE. Reprinted with permission.)

estimation (MLE) of the location of the target is given by

$$\hat{\mathbf{x}} = \underset{\mathbf{x}}{\operatorname{argmin}} F(\mathbf{x}, \boldsymbol{\varphi}) \tag{2.270}$$

where

$$F(\mathbf{x}, \boldsymbol{\varphi}) = \frac{1}{2}[\mathbf{g}(\mathbf{x}) - \boldsymbol{\varphi}]^{\mathrm{T}} \mathbf{S}^{-1} [\mathbf{g}(\mathbf{x}) - \boldsymbol{\varphi}] \tag{2.271}$$

is a cost function. The components of the cost function are

$$\mathbf{g}(\mathbf{x}) = [g_1(\mathbf{x}) \quad \cdots \quad g_M(\mathbf{x})]^{\mathrm{T}} \tag{2.272}$$

In two dimensions,

$$\mathbf{x} = [x_{\mathrm{T}} \quad y_{\mathrm{T}}]^{\mathrm{T}} \tag{2.273}$$

and the bearing measurements are given by (see Figure 2.35)

$$\boldsymbol{\varphi} = [\phi_1 \quad \phi_2 \quad \cdots \quad \phi_M]^{\mathrm{T}} \tag{2.274}$$

The components of $\mathbf{g}(\mathbf{x})$ are given by

$$g_i(\mathbf{x}) = \tan^{-1}\left(\frac{\Delta_{y_i}}{\Delta_{x_i}}\right) \tag{2.275}$$

where

$$\begin{aligned} \Delta_{x_i} &= x_{\mathrm{T}} - x_i \\ \Delta_{y_i} &= y_{\mathrm{T}} - y_i \end{aligned} \tag{2.276}$$

Finally, $\mathbf{S} = \operatorname{diag}\left(\sigma_1^2, \sigma_2^2, \cdots, \sigma_M^2\right)$ is the $M \times M$ covariance matrix of the LOB observations.

M is the number of observations and there can be multiple observations per sensor. However, the number of observations from all sensors is assumed to be the same.

The LOB observations are corrupted with AWGN given by

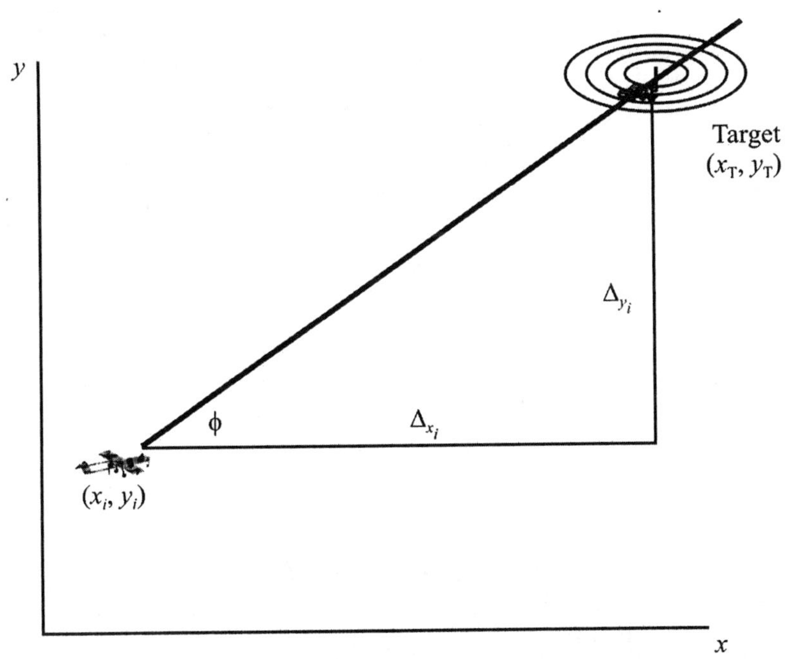

Figure 2.35 MLE LOBs.

$$\boldsymbol{\delta\varphi} = [\delta\phi_1 \quad \delta\phi \quad \cdots \quad \delta\phi_M]^{\mathrm{T}} \tag{2.277}$$

When the true LOBs are given by $\boldsymbol{\varphi}_0$, the observed LOBs are given by

$$\boldsymbol{\varphi} = \boldsymbol{\varphi}_0 + \boldsymbol{\delta\varphi} \tag{2.278}$$

Expression (2.271) can be put in the form

$$F(\mathbf{x}, \boldsymbol{\varphi}) = \frac{1}{2}\mathbf{f}^{\mathrm{T}}\mathbf{S}^{-1}\mathbf{f} = \frac{1}{2}\sum_{i=1}^{M}\frac{f_i^2}{\sigma_i^2} \tag{2.279}$$

where

$$\mathbf{f} = [f_1 \quad f_2 \quad \cdots \quad f_M]^{\mathrm{T}} = \mathbf{g}(\mathbf{x}) - \boldsymbol{\varphi} \tag{2.280}$$

The solution of (2.270), which is a nonlinear equation, can be found numerically by the Newton-Gauss method, which iterates to a solution with

$$\hat{\mathbf{x}}_{m+1} = \hat{\mathbf{x}}_m + [\mathbf{g}_x^{\mathrm{T}}\mathbf{S}^{-1}\mathbf{g}_x]^{\mathrm{T}}\mathbf{g}_x^{\mathrm{T}}\mathbf{S}^{-1}[\boldsymbol{\varphi} - \mathbf{g}(\hat{\mathbf{x}}_m)] \tag{2.281}$$

where $\mathbf{g}_x = \partial\mathbf{g}/\partial\mathbf{x}$ evaluated at the true target position, yielding

$$\mathbf{g}_x = \begin{bmatrix} -\dfrac{\Delta_{y_1}}{r_1^2} & -\dfrac{\Delta_{y_2}}{r_2^2} & \cdots & -\dfrac{\Delta_{y_M}}{r_M^2} \\ \dfrac{\Delta_{x_1}}{r_1^2} & \dfrac{\Delta_{x_2}}{r_2^2} & \cdots & \dfrac{\Delta_{x_M}}{r_M^2} \end{bmatrix} \tag{2.282}$$

with

$$r_i^2 = \Delta_{x_i}^2 + \Delta_{y_i}^2 \tag{2.283}$$

■

The two MLE techniques analyzed by [20] are the original Stansfield algorithm [21] and the true MLE algorithm as given by (2.279) to (2.281). The analysis presented in this section follows [22] closely.

One of the first algorithms developed for the purpose of calculating the location of an emitter based on multiple lines of bearing was due to Stansfield [21]. In that algorithm it is assumed that the bearing errors of the EW systems are normally distributed. The joint probability density function of multiple lines of bearings is then a multivariate Gaussian probability density function. A maximum likelihood estimator for the PF ensues by maximizing the exponent in the equation for the joint probability density function (which will minimize the total probability of error because the exponent is negative).

Stansfield's original technique for calculating the PF of an emitter is probably the first example of maximum likelihood-like estimation applied to the PF problem. The geometry involved is illustrated in Figure 2.36. As described in [21–23], the goal is to minimize the expression for the joint probability of miss given as a function of the miss distances given by

$$P(d_1, d_2, \cdots, d_N) = \frac{1}{(2\pi)^{M/2}\sum_{i=1}^{M}\sigma_{p_i}} \exp\left(-\frac{1}{2}\sum_{i=1}^{M}\frac{d_i^2}{\sigma_{p_i}^2}\right) \tag{2.284}$$

This expression is minimized when the argument of exp is maximized. The result is the PF that is the most likely location of the target.

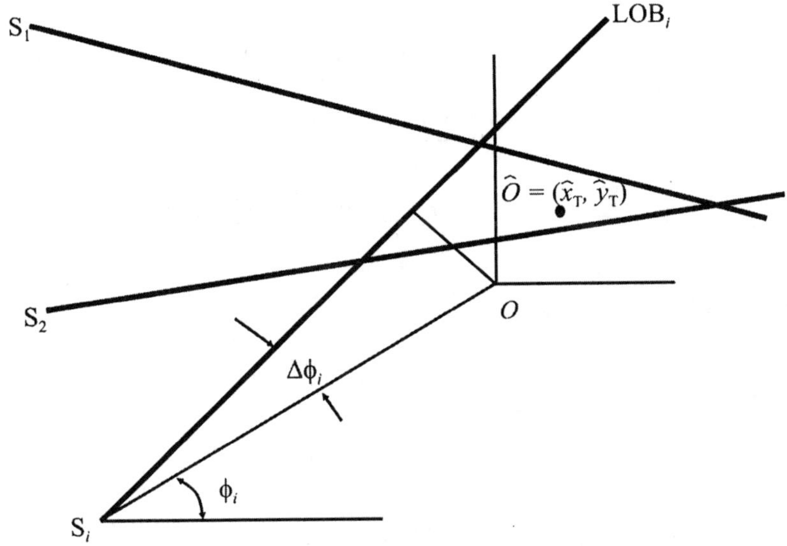

Figure 2.36 Stansfield's PF geometry.

The MLE cost function from (2.279) is given by

$$F_{\text{ML}}(\mathbf{x}, \mathbf{\Phi}) = \frac{1}{2}\sum_{i=1}^{N}\frac{f_i^2}{\sigma_i^2} \tag{2.285}$$

where $\mathbf{x} = (x_T, y_T)$ is the target location to be estimated. The Stansfield fix algorithm is an example of an MLE-like algorithm, where it is assumed that the errors in bearing measurement are small. In (2.285), the f_i represent the difference between the measured bearings and the bearings corresponding to a target at the estimated position given by $\hat{\mathbf{x}} = (\hat{x}_T, \hat{y}_T)$, as shown in Figure 2.36. Thus, $f_i = \Delta\phi_i$. Stansfield's approach assumes that the bearing errors are small, so that $\sin(\Delta\phi) \approx \Delta\phi$. The cost function is thus given by

$$F_{\text{ST}}(\mathbf{x}, \mathbf{\Phi}) = \frac{1}{2}\sum_{i=1}^{M}\frac{\sin^2 \Delta\phi_i}{\sigma_i^2} \tag{2.286}$$

However [20],

$$\sin f_i = \sin\left[\tan^{-1}\left(\frac{\Delta_{y_i}}{\Delta_{x_i}}\right) - \phi_i\right]$$
$$= \frac{\Delta_{y_i}\cos\phi_i - \Delta_{x_i}\sin\phi_i}{r_i}$$
$$= \frac{(y_\text{T} - y_i)\cos\phi_i - (x_\text{T} - x_i)\sin\phi_i}{r_i}$$

where

$$r_i^2 = \Delta_{x_i}^2 + \Delta_{y_i}^2, \qquad i = 1, 2, \ldots, K$$

Then

$$F_\text{ST}(\mathbf{x}, \mathbf{\Phi}) = \frac{1}{2}\sum_{i=1}^{M}\frac{[(y_\text{T} - y_i)\cos\phi_i - (x_\text{T} - x_i)\sin\phi_i]^2}{r_i^2\sigma_i^2}$$
$$= \frac{1}{2}(\mathbf{Ax} - \mathbf{b})^\text{T}\mathbf{D}^{-1}\mathbf{S}^{-1}(\mathbf{Ax} - \mathbf{b}) \qquad (2.287)$$

where

$$\mathbf{A} = \begin{bmatrix} \sin\phi_1 & -\cos\phi_1 \\ \vdots & \vdots \\ \sin\phi_M & -\cos\phi_M \end{bmatrix}$$
$$\mathbf{b} = \begin{bmatrix} x_1\sin\phi_1 - y_1\cos\phi_1 \\ \vdots \\ x_M\sin\phi_M - y_M\cos\phi_M \end{bmatrix}$$
$$\mathbf{D} = \text{diag}(r_1^2, r_2^2, \ldots, r_N^2)$$

The value of $\mathbf{x}_\text{ST}$ that minimizes (2.287) is given by (2.17) with the weighting matrix as $\mathbf{W} = \mathbf{D}^{-1}\mathbf{S}^{-1}$, which yields

$$\hat{\mathbf{x}}_\text{ST} = (\mathbf{A}^\text{T}\mathbf{D}^{-1}\mathbf{S}^{-1}\mathbf{A})^{-1}\mathbf{A}^\text{T}\mathbf{D}^{-1}\mathbf{S}^{-1}\mathbf{b} \qquad (2.288)$$

2.8.2.1 Bias and Variance of the Stansfield PF Estimator

The bias and variance of the Stansfield PF estimator are examined in this section. As in Section 2.7.1, it is assumed that there are N sensors, each making P observations so that $M = NP$ is the total number of observations. It is shown in [20] that the bias of the Stansfield PF estimator is given by

$$\mathcal{E}\{\delta\mathbf{x}\} \approx -\frac{1}{P}\tilde{\mathbf{C}}\mathbf{h} \tag{2.289}$$

where

$$\mathbf{h} = \sum_{k=1}^{N} \frac{1}{r_k^2} \left(\begin{array}{c} \dfrac{1}{\sigma_k^2 r_k^4} \begin{bmatrix} 2\Delta_{x_k}\Delta_{y_k} & \Delta_{y_k}^2 - \Delta_{x_k}^2 \\ \Delta_{y_k}^2 - \Delta_{x_k}^2 & -2\Delta_{x_k}\Delta_{y_k} \end{bmatrix} \\ \times \tilde{\mathbf{C}} \begin{bmatrix} -\Delta_{y_k} \\ \Delta_{x_k} \end{bmatrix} + P \begin{bmatrix} \Delta_{x_k} \\ \Delta_{y_k} \end{bmatrix} \end{array} \right) \tag{2.290}$$

and

$$\tilde{\mathbf{C}}^{-1} = \tilde{\mathbf{g}}_x^{\mathrm{T}} \mathbf{S}^{-1} \tilde{\mathbf{g}}_x = \sum_{k=1}^{N} \frac{1}{\sigma_k^2 r_k^4} \begin{bmatrix} \Delta_{y_k}^2 & -\Delta_{x_k}\Delta_{y_k} \\ -\Delta_{x_k}\Delta_{y_k} & \Delta_{x_k}^2 \end{bmatrix} \tag{2.291}$$

The tilde indicates that the matrix is evaluated at $P = 1$.

Thus,

$$\lim_{P\to\infty} \mathcal{E}\{\delta\mathbf{x}\} \approx -\tilde{\mathbf{C}} \sum_{k=1}^{N} \begin{bmatrix} \dfrac{\Delta_{x_k}}{r_k^2} \\ \dfrac{\Delta_{y_k}}{r_k^2} \end{bmatrix} \tag{2.292}$$

which indicates that the Stansfield algorithm produces biased results, since, even for very large numbers of samples, M, the bias does not disappear.

The covariance of this estimator is given by [20]

$$\mathbf{Cov}(\delta\mathbf{x}) \approx \frac{1}{P}\tilde{\mathbf{C}}\tilde{\mathbf{H}}\tilde{\mathbf{S}}^{-1}\tilde{\mathbf{H}}^{\mathrm{T}}\tilde{\mathbf{C}}\Big|_{(\mathbf{x}_b,\mathbf{\Phi}_0)} \tag{2.293}$$

where

$$\tilde{\mathbf{H}} = \sum_{k=1}^{N}\frac{1}{r_k^2}\begin{bmatrix}\Delta_{x_k}\sin 2\phi_k - \Delta_{y_k}\cos 2\phi_k \\ -\Delta_{x_k}\sin 2\phi_k - \Delta_{y_k}\cos 2\phi_k\end{bmatrix}\tilde{\mathbf{e}}_k^{\mathrm{T}}$$

$$\mathbf{x}_b = \mathbf{x}_0 + \mathcal{E}\{\delta\mathbf{x}\}$$

$\tilde{\mathbf{e}}_k^{\mathrm{T}}$ is the kth column of the $N \times N$ identity matrix $\mathbf{I}_N$

Because the Stansfield PF estimator is biased, the *Cramer-Rao lower bound* (CRLB) cannot be used as an estimate of performance, since the CRLB only applies to unbiased estimators. If an unbiased PF estimator is sought, the Stansfield algorithm is probably a poor choice.

Example: Suppose there are two sensor locations at ($-a$, 0, 0) and (a, 0, 0) and a target located at (0, b, 0), as shown in Figure 2.37. Each sensor collects P observations with variance σ^2, assumed to be the same for all sensors. Therefore, $N = 2$, $M = 2P$. In this case, from (2.282),

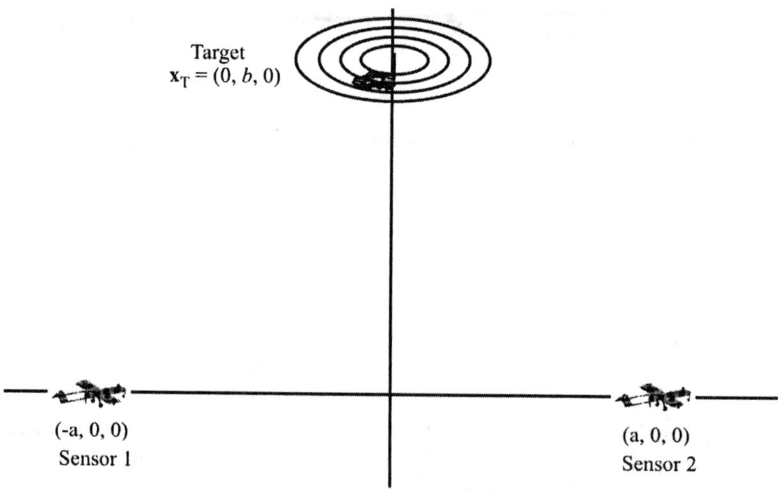

Figure 2.37 Geometry for MLE algorithm example.

$$\mathbf{g}_{\mathrm{x}}^{\mathrm{T}} = \begin{bmatrix} \overbrace{\begin{matrix} \dfrac{-b}{a^2+b^2} & \dfrac{-a}{a^2+b^2} & \cdots & \dfrac{-b}{a^2+b^2} & \dfrac{-a}{a^2+b^2} \\ \dfrac{-b}{a^2+b^2} & \dfrac{a}{a^2+b^2} & \cdots & \dfrac{-b}{a^2+b^2} & \dfrac{a}{a^2+b^2} \end{matrix}}^{M=2P \text{ times}} \end{bmatrix} \tag{2.294}$$

and

$$\mathbf{S}^{-1} = \begin{bmatrix} \dfrac{1}{\sigma^2} & 0 & \cdots & 0 \\ 0 & \dfrac{1}{\sigma^2} & \cdots & 0 \\ \vdots & \vdots & \ddots & \vdots \\ 0 & 0 & \cdots & \dfrac{1}{\sigma^2} \end{bmatrix} \tag{2.295}$$

with matrix $\mathbf{C}^{-1}$ given as

$$\mathbf{C}^{-1} = \mathbf{g}_{\mathrm{x}}^{\mathrm{T}} \mathbf{S}^{-1} \mathbf{g}_{\mathrm{x}} = \begin{bmatrix} \dfrac{-b}{a^2+b^2} & \dfrac{-a}{a^2+b^2} & \cdots & \dfrac{-b}{a^2+b^2} & \dfrac{-a}{a^2+b^2} \\ \dfrac{-b}{a^2+b^2} & \dfrac{a}{a^2+b^2} & \cdots & \dfrac{-b}{a^2+b^2} & \dfrac{a}{a^2+b^2} \end{bmatrix}$$
$$\times \begin{bmatrix} \dfrac{1}{\sigma^2} & 0 & \cdots & 0 \\ 0 & \dfrac{1}{\sigma^2} & \cdots & 0 \\ \vdots & \vdots & \ddots & \vdots \\ 0 & 0 & \cdots & \dfrac{1}{\sigma^2} \end{bmatrix} \begin{bmatrix} \dfrac{-b}{a^2+b^2} & \dfrac{-b}{a^2+b^2} \\ \dfrac{-a}{a^2+b^2} & \dfrac{a}{a^2+b^2} \\ \vdots & \vdots \\ \dfrac{-b}{a^2+b^2} & \dfrac{-b}{a^2+b^2} \\ \dfrac{-a}{a^2+b^2} & \dfrac{a}{a^2+b^2} \end{bmatrix} \tag{2.296}$$

which reduces to

$$\mathbf{C} = \frac{\sigma^2 (b^2 + a^2)^2}{2P} \begin{bmatrix} \frac{1}{b^2} & 0 \\ 0 & \frac{1}{a^2} \end{bmatrix} \tag{2.297}$$

From (2.289) and (2.290),

$$\mathcal{E}\{\delta\mathbf{x}\} \approx -\frac{\sigma^2 b(b^2 + a^2)}{a^2} \begin{bmatrix} 0 \\ 1 - \frac{3}{P} + \frac{a^2}{2Pb^2} \end{bmatrix} \tag{2.298}$$

so that

$$\lim_{P \to \infty} \mathcal{E}\{\delta\mathbf{x}\} = -\frac{\sigma^2 b(b^2 + a^2)}{a^2} \begin{bmatrix} 0 \\ 1 \end{bmatrix} \tag{2.299}$$

In the x direction, the limit converges to zero, forming an unbiased estimator. The same cannot be said for the y direction. There is a residual bias.

The results of this analysis for the two estimators are illustrated in Figure 2.38, with the MLE as solid lines and the Stansfield estimator as dashed lines. The total RMS error continues to decrease for the MLE as the number of measurements increases, while that for the Stansfield estimator levels after about 10 measurements. The y bias for the MLE approaches zero, while the y bias for the Stansfield estimator settles to a relatively large negative value. The x and y standard deviations for both estimators are approximately the same.

■

2.9 Multiple Sample Correlation

Fu, Vian, and Grose developed an algorithm for position fixing that utilizes the intersection of the LOB fans, defined as the measured LOB plus and minus the maximum error in the sensor [24]. Using the maximum error guarantees that the emitter is within the resulting area. If 1σ or other statistical quantity is used instead of the maximum error, this guarantee cannot be assumed. However, the EEP can be calculated with a certain probability. The basic idea is illustrated in Figure 2.39. Successive LOBs are obtained from a single sensor or simultaneous LOBs from multiple sensors. The area on the surface of the Earth defined by the sides of these

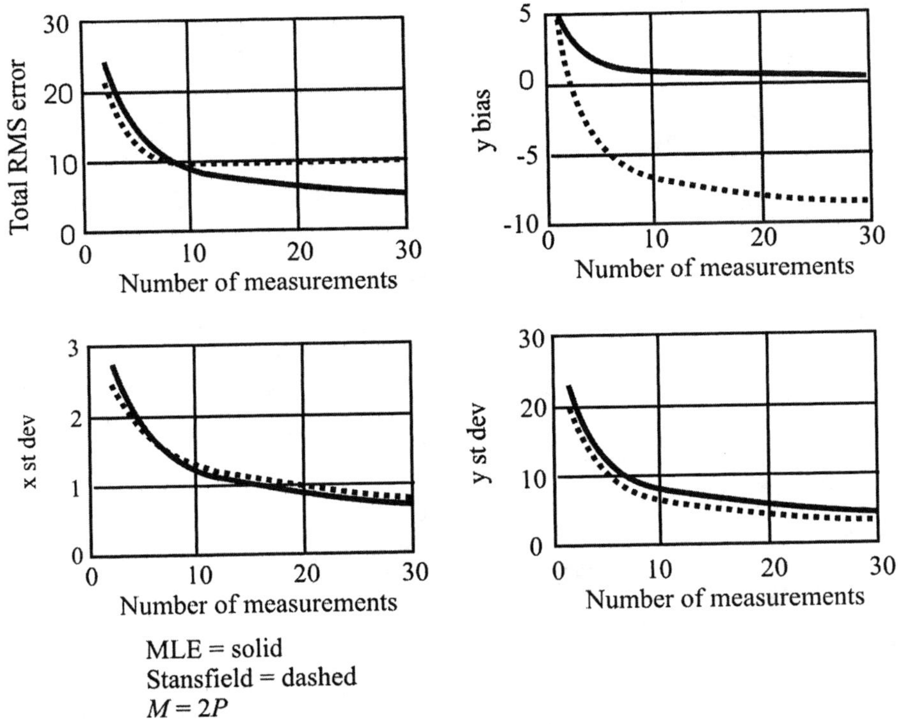

Figure 2.38 MLE example results. (*Source:* [20], © 1992 IEEE. Reprinted with permission.)

fans is the estimate of the region within which the target lies. The steps in the algorithm are given in Figure 2.40 [24].

Example: The reference describes a test case, the flight paths of which are shown in Figure 2.41. Two flight paths are described, the first a straight line along the x-axis and the second a curvilinear path with a constant turn rate of 1° per sec. The bearing measurement error was assumed to be ±1°.

Figure 2.42 illustrates the results. Flight path 2 ultimately yielded more accurate PFs, primarily because of how close to the target the sensor traveled.

A comparison to the Kalman filter described by Spingarn [14] was also described based on Monte Carlo simulations of both algorithms. The results are shown in Table 2.4.

■

The idea of using maximum error is appealing because it guarantees the emitter is within the error bounds. Unfortunately, it is difficult to guarantee that

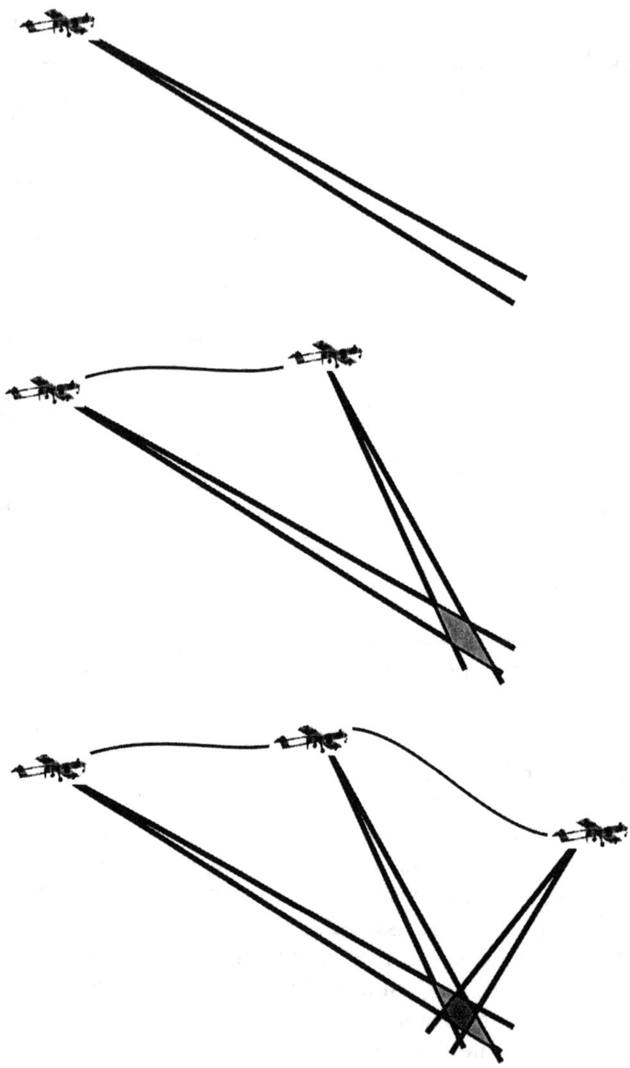

Figure 2.39 Multiple sample correlation position fixing.

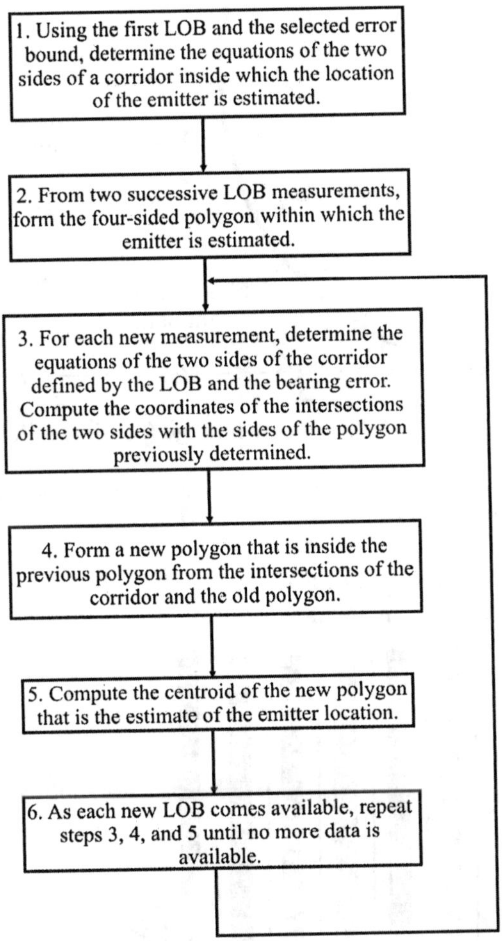

Figure 2.40 Multiple sample correlation algorithm.

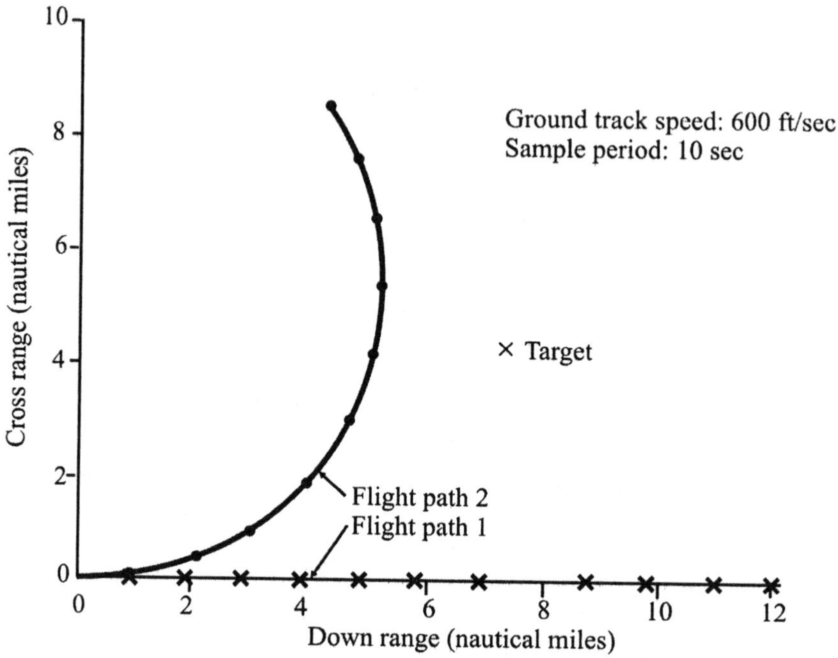

Figure 2.41 Flight paths for the multiple sample correlation. (*Source:* [24], © 1988 IEEE. Reprinted with permission.)

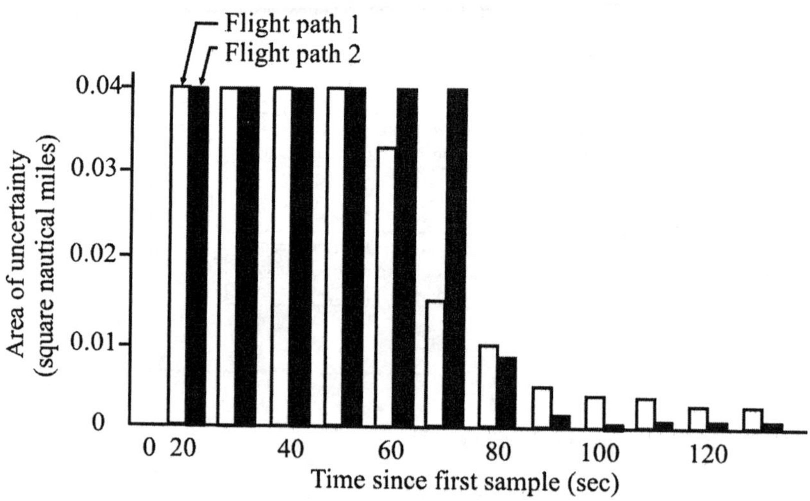

Figure 2.42 Emitter area of uncertainty. For the first 40 sec for flight path 1 and 60 sec for flight path 2, the areas of uncertainty are off the scale (larger than 0.04 nautical miles). (*Source:* [24], © 1988 IEEE. Reprinted with permission.)

Table 2.4 Emitter Location Error

Standard deviation (degrees)	Maximum error (degrees)	Percent with maximum error	EKF error (feet)	MSC error (feet)
1	1	68.3	5,477	1,633
1	2	95.6	5,477	3,267
1	3	99.7	5,477	4,900

Source: [24]. *Multiple sample correlation.

there is some maximum error associated with an LOB measurement device. Operational factors are frequently the dominant source of LOB error, especially for sensors close to or on the ground, where reflections off buildings, the ground, passing vehicles, and so forth can cause LOBs that are off by a full half circle. In those cases where multipath reflections are not likely to cause problems, such as satellite sensors or high-flying UAVs, the approach is more viable.

2.10 Bearing-Only Target Motion Analysis

It is possible to track moving targets with bearing-only data (called *bearing-only* TMA). In TMA it is desirable to determine the location and velocity of a moving target, in general. For a set of instantaneous LOBs, the velocity cannot be determined, which is a special case of TMA—the target is not moving.

TMA is based on the scenario illustrated in Figure 2.35. Only two dimensions will be included here for simplicity. The target, located at (x_T, y_T), is moving with a constant velocity $(v_{\text{T}x}, v_{\text{T}y})$. The state vector associated with the target at time i is thus $\mathbf{x}_\text{T}(t_i)=\begin{bmatrix}x_\text{T}(t_i) & y_\text{T}(t_i) & v_{\text{T}_x}(t_i) & v_{\text{T}_y}(t_i)\end{bmatrix}^\text{T}$. The sensor state vector is defined similarly as $\mathbf{x}_\text{R}(t_i)=\begin{bmatrix}x_\text{R}(t_i) & y_\text{R}(t_i) & v_{\text{R}_x}(t_i) & v_{\text{R}_y}(t_i)\end{bmatrix}^\text{T}$. The velocity of the sensor is not necessarily constant, so it is therefore a function of t_i. The relative target-sensor state vector is given by

$$\mathbf{x}(t_i)=\mathbf{x}_\text{T}(t_i)-\mathbf{x}_\text{R}(t_i)=\begin{bmatrix}\Delta_x(t_i) & \Delta_y(t_i) & v_x(t_i) & v_y(t_i)\end{bmatrix}^\text{T} \tag{2.300}$$

The discrete time equation describing this dynamic system is given by

$$\mathbf{x}(t_i)=\mathbf{\Phi}(t_i,t_{i-1})\mathbf{x}(t_i)+\mathbf{u}(t_i) \tag{2.301}$$

where the state transition matrix is given by

$$\boldsymbol{\Phi}(t_i, t_{i-1}) = \begin{bmatrix} \mathbf{I} & (t_i - t_{i-1})\mathbf{I} \\ \mathbf{Z} & \mathbf{I} \end{bmatrix} \tag{2.302}$$

$$\mathbf{I} = \begin{bmatrix} 1 & 0 \\ 0 & 1 \end{bmatrix} \tag{2.303}$$

$$\mathbf{Z} = \begin{bmatrix} 0 & 0 \\ 0 & 0 \end{bmatrix} \tag{2.304}$$

and t_i is the ith sample time while t_{i-1} is the i–1st sample time. The term $\mathbf{u}(t_i) = [0 \quad 0 \quad u_x(t_i) \quad u_y(t_i)]^{\mathrm{T}}$ in (2.301) accounts for the effects of acceleration of the sensor.

The measurements in this system consist of bearings collected at the sensor referenced to the positive x-axis and measured counterclockwise-positive, corrupted by noise as

$$\phi_i = \theta_i + \eta_i \tag{2.305}$$

where θ_i is the true bearing to the target given by

$$\theta_i = \tan^{-1} \frac{\Delta_x(t_i)}{\Delta_y(t_i)} \tag{2.306}$$

and $\eta_i \sim \mathcal{N}(0, \sigma_\eta^2)$, $\eta_i \in (-\pi, \pi]$.

These equations describe a nonlinear process. They can be converted into a linear problem by defining a new measurement given by

$$\alpha_i = \mathbf{h}^{\mathrm{T}}(\phi_i)\mathbf{x}_{\mathrm{R}}(t_i) \tag{2.307}$$

and

$$\alpha_i = \mathbf{h}^{\mathrm{T}}(\phi_i)\mathbf{x}_{\mathrm{T}} + r_i \sin \eta_i \tag{2.308}$$

where

$$r_i = \sqrt{\Delta_x^2(t_i) + \Delta_y^2(t_i)} \tag{2.309}$$

and

$$\mathbf{h}^{\mathrm{T}}(\phi_i) = [\cos\phi_i \quad -\sin\phi_i \quad 0 \quad 0] \tag{2.310}$$

Since the noise is assumed to be Gaussian and additive, the series of bearing measurements given by $\mathbf{\Phi} = [\phi_1 \quad \phi_2 \quad \cdots \quad \phi_i]$ have the likelihood function

$$P_{\Phi|\mathbf{x}} = \frac{1}{\sqrt{(2\pi)^i \det(\mathbf{W})}} \exp\left\{-\frac{1}{2}[\mathbf{\Phi} - \mathbf{\Phi}(\mathbf{x})]^{\mathrm{T}} \mathbf{W}^{-1} [\mathbf{\Phi} - \mathbf{\Phi}(\mathbf{x})]\right\} \tag{2.311}$$

where $\Phi(\mathbf{x})$ is the vector of correct bearings and

$$\mathbf{W} = \mathrm{diag}(\sigma_{\alpha_i}^2) \tag{2.312}$$

The MLE is given by the solution of

$$\frac{\partial \ln P\mathbf{\Phi}|\mathbf{x}}{\partial \mathbf{x}} = 0 \tag{2.313}$$

Using (2.311), this equates to

$$\left[\frac{\partial \mathbf{\Phi}(\mathbf{x})}{\partial \mathbf{x}}\right]^{\mathrm{T}} \mathbf{W}^{-1}[\mathbf{\Phi} - \mathbf{\Phi}(\mathbf{x})] = \mathbf{0} \tag{2.314}$$

Carrying out the math finally yields the iteration equation

$$\mathbf{x}_{k+1} = \mathbf{x}_k - \lambda_k [\hat{\mathbf{\Gamma}}_{\mathbf{\Phi}}^{\mathrm{T}} \hat{\mathbf{R}}^{-1} \mathbf{W}^{-1} \hat{\mathbf{R}}^{-1} \hat{\mathbf{\Gamma}}_{\mathbf{\Phi}}]^{-1} \hat{\mathbf{\Gamma}}_{\mathbf{\Phi}}^{\mathrm{T}} \hat{\mathbf{R}}^{-1} \mathbf{W}^{-1} [\mathbf{\Phi} - \hat{\mathbf{\Phi}}] \tag{2.315}$$

where

$$\hat{\mathbf{\Phi}} = \mathbf{\Phi}(\hat{\mathbf{x}})$$

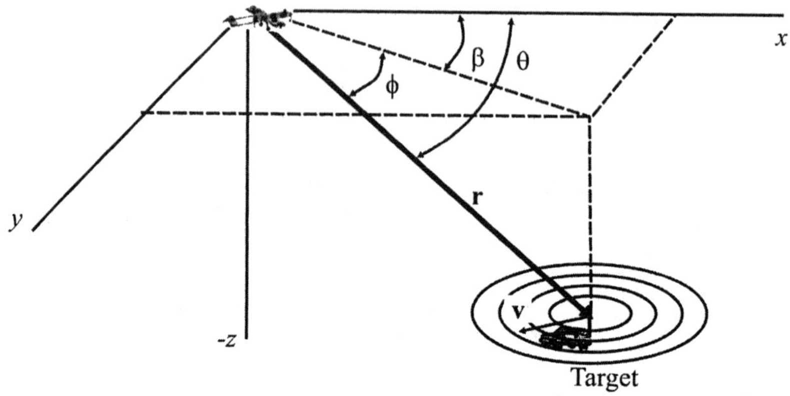

Figure 2.43 Geometry to establish observability requirements.

$$\mathbf{\Gamma}_\Phi = \begin{bmatrix} \cos\hat{\phi}_1 & -\sin\hat{\phi}_1 & (t_1 - t_m)\cos\hat{\phi}_1 & (t_1 - t_m)\sin\hat{\phi}_1 \\ \cos\hat{\phi}_2 & -\sin\hat{\phi}_2 & (t_1 - t_m)\cos\hat{\phi}_2 & (t_1 - t_m)\sin\hat{\phi}_2 \\ \vdots & & & \\ \cos\hat{\phi}_i & -\sin\hat{\phi}_i & (t_1 - t_m)\cos\hat{\phi}_i & (t_1 - t_m)\sin\hat{\phi}_i \end{bmatrix}$$

$$\hat{\mathbf{R}} = \text{diag}(r_i)$$

Parameter λ_k is the step size at step k, and t_m is the reference time when the time scale started.

The general requirement for being able to perform bearing-only TMA in terms of observability requirements was established by Hammel and Aidala [25] in an underwater sonar context. Assuming a single moving sensor system, then with the notation as indicated in Figure 2.43, the requirements to be able to determine the location and velocity of the target (to be able to *observe* the target[2]) are given by

---

[2] A dynamical system described by the normal form equations

$$\dot{\mathbf{x}} = \mathbf{Ax} + \mathbf{Bv} \qquad \mathbf{y} = \mathbf{Cx} + \mathbf{Dv} + \mathbf{E}\dot{\mathbf{v}}$$

with n states in vector $\mathbf{x}$, m inputs in vector $\mathbf{v}$, and p outputs in vector $\mathbf{y}$, with $\mathbf{A}_{n\times n}$ the state transition matrix, $\mathbf{B}_{n\times m}$, $\mathbf{C}_{p\times n}$, $\mathbf{D}_{p\times m}$, and $\mathbf{E}_{p\times m}$, is *controllable* if, given any initial state $\mathbf{x}_0$ at t_0, a set of inputs exists that can drive the system to the zero state at t_1. This system is *observable* if, observing the outputs over (t_0, t_1), the initial state of the system at t_0 can be determined.

$$\det(\mathbf{H}^{\mathrm{T}}\mathbf{H}) \neq 0 \tag{2.316}$$

where **H** is the matrix given by

$$\mathbf{H} = \begin{bmatrix} \mathbf{I} & \mathbf{h}_0 & \mathbf{Z} & \mathbf{0} \\ \mathbf{Z} & \mathbf{h}_1 & \mathbf{I} & \mathbf{h}_0 \\ \mathbf{Z} & \mathbf{h}_2 & \mathbf{Z} & 2\mathbf{h}_1 \\ \mathbf{Z} & \mathbf{h}_3 & \mathbf{Z} & 3\mathbf{h}_2 \\ \mathbf{Z} & \mathbf{h}_4 & \mathbf{Z} & 4\mathbf{h}_3 \\ \mathbf{Z} & \mathbf{h}_5 & \mathbf{Z} & 5\mathbf{h}_4 \end{bmatrix} \tag{2.317}$$

with **I** and **Z** given by (2.303) and (2.304) and

$$\mathbf{h}_i = \begin{bmatrix} f_i \\ g_i \end{bmatrix} = -\begin{bmatrix} \dfrac{d^i(\sin\beta\cot\phi)}{dt^i} \\ \dfrac{d^i(\cos\beta\cot\phi)}{dt^i} \end{bmatrix}, \qquad i = 0,1,\ldots,5$$

Evaluation of the determinate expressed by (2.316) yields the equivalent requirement for observability as

$$\det(\mathbf{H}^{\mathrm{T}}\mathbf{H}) = \frac{1}{2}\sum_{i=1}^{4}\sum_{j=1}^{4}\left\{ \begin{array}{l} [(i+1)f_i f_{j+1} - (j+1)f_i f_{i+1}]^2 \\ +2[(i+1)f_i g_{i+1} - (j+1)g_i f_{i+1}]^2 \\ +[(i+1)g_i g_{i+1} - (j+1)g_j g_{i+1}]_2 \end{array} \right\} \tag{2.318}$$

Requirement (2.316) is satisfied if at least one of the terms in the double summation in (2.318) does not vanish.

Sensor or target motion is required in order to evaluate the velocity terms. If more than one sensor is available so that multiple instantaneous LOBs are available then geoposition determination is possible from a single set of LOBs, but velocity still cannot be determined.

Note that if the sensor as well as the target is moving at a constant velocity, the system is unobservable [25]. When there is only one moving sensor involved, this requirement dictates that the sensor must perform some sort of maneuver so that the velocity vector of the sensor is changed. (It is tacitly assumed that the velocity vector of the target cannot be affected by the sensor.) Changing the velocity vector implies an acceleration of the sensor. Thus, the fundamental requirement is for the sensor to accelerate in some fashion in order for the target to be observable. It should be noted that some forms of acceleration are better than others.

The portions of the sensor path that are defined by different velocity vectors are referred to as *legs*. The target is unobservable in the first leg of the sensor path when there is only one sensor involved. If the motion of the single sensor is co-linear[3] with the motion of the target, the system is not observable as well. Acceleration is required in this case as well that changes the direction of the velocity vector.

It is relatively easy and fast to maneuver an airborne sensor observing a target on the ground or on the sea. For surface sensors, such as a ship with sonar, this maneuvering can take considerably more time. The same can be said for sub-surface boats.

2.11 Sources of Error in Triangulation

There are several sources of error that can enter into the PF estimation by triangulation. Two have already been discussed at length: noise and measurement errors. There are others, however. Some of these are discussed in this section.

2.11.1 Geometric Dilution of Precision in Triangulation

As explained in [26], the geometry of the baseline of sensors has an effect on the accuracy of the geopositions computed. Generally, the farther from the baseline, the less accurate the PF estimate will be. This is illustrated in Figure 2.44. The sensors shown at the top and bottom of Figure 2.44 have the same LOB accuracy, yet the area within which the target lies is substantially larger the farther it is from the baseline of the sensors. As can be seen, the farther from the baseline the target is, the LOBs become more and more parallel. This effect is called *geometric dilution of precision* (GDOP). These effects are quantified in the example shown in Figure 2.45 [27], where CEP/σ normalized by d, the length of the baseline, is plotted versus x and y, also normalized.

GDOP also has degradation effects when the target is moved off the perpendicular bisector of the baseline. These effects are also illustrated in Figure 2.45.

---

[3] A system moving with velocity vector $\mathbf{v}_o(t)$ is moving *colinearly* with another system moving with velocity $\mathbf{v}_T(t)$ if $\mathbf{v}_o(t) = \mathbf{k}\mathbf{v}_T(t)$ for some constant vector $\mathbf{k}$. Expressed as 3D vectors, this means that the two systems are moving in the same direction.

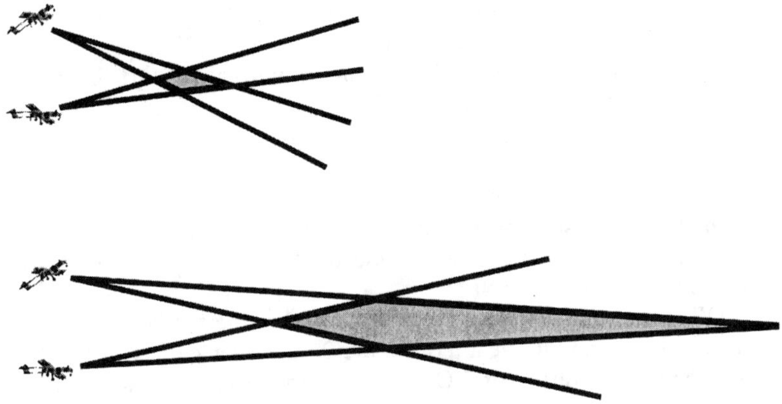

Figure 2.44 Distance effects on PF calculation accuracy.

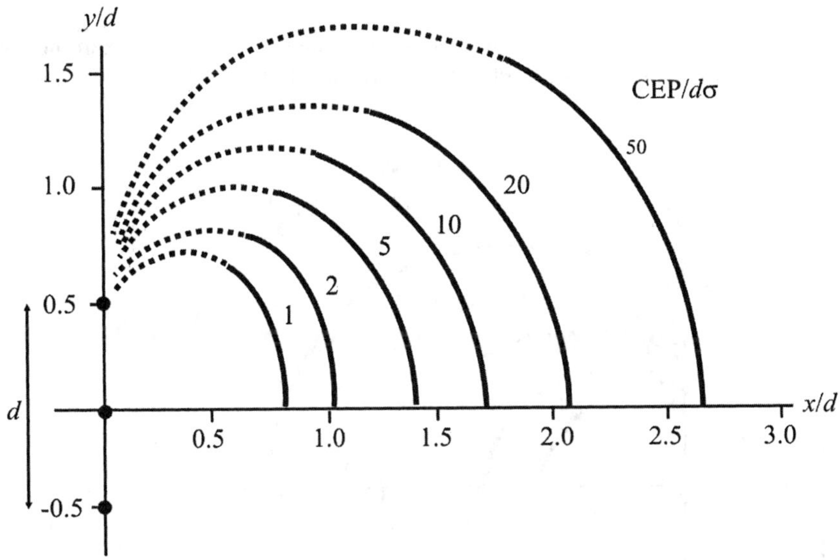

Figure 2.45 The effects of GDOP when a linear baseline is employed. (*Source:* [27], © 1984 IEEE. Reprinted with permission.)

The shape of the baseline also has an effect. Off the end of a baseline between any two sensors, the LOBs become linearly dependent (they are colinear) and the fix accuracy explodes. This is illustrated in Figure 2.46 [27]. The effect is seen here by the range over which the CEP is some finite value. Off the ends of the baselines the CEP increases without bound.

2.11.2 LOB Error

The constituent LOBs used for the PF estimation are never error free. Noise and systematic errors are always present. Some of the systematic error can be, and usually is, removed by calibration.

The LOB error contribution increases the farther the sensor array is from the target. This can be seen with the aid of Figure 2.44. The accuracy of LOB systems is usually specified in terms of LOB angular error. The LOBs in Figure 2.44 represent the 1−σ limits and are the same for both parts of the figure. Notice that the area in the error zone in the lower part of the figure is much larger than that at the top. This is because in the lower case, the PF is farther from the sensor baseline.

2.11.3 Effects of Bias on Bearing-Only PF

The effects of bias errors on bearing-only PF calculations were examined by Gavish and Fogel [28]. The geometry is depicted in Figure 2.47. It is assumed that

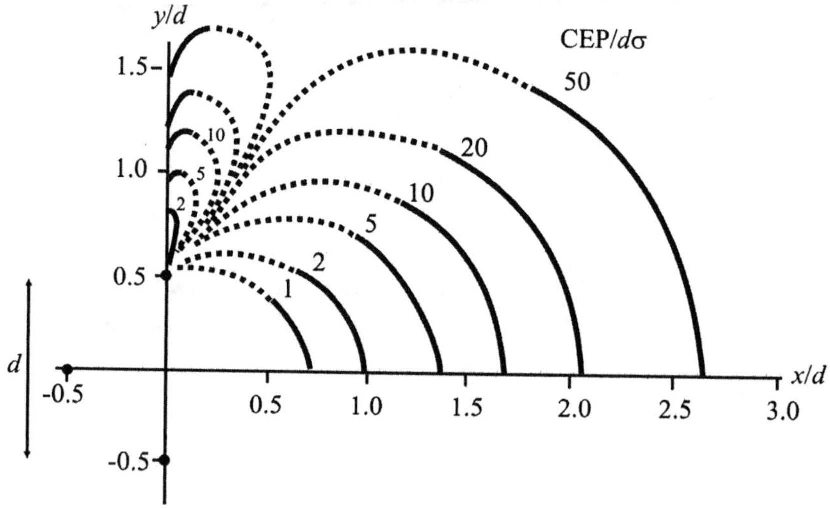

Figure 2.46 GDOP V-baseline. (*Source:* [27], © 1984 IEEE. Reprinted with permission.)

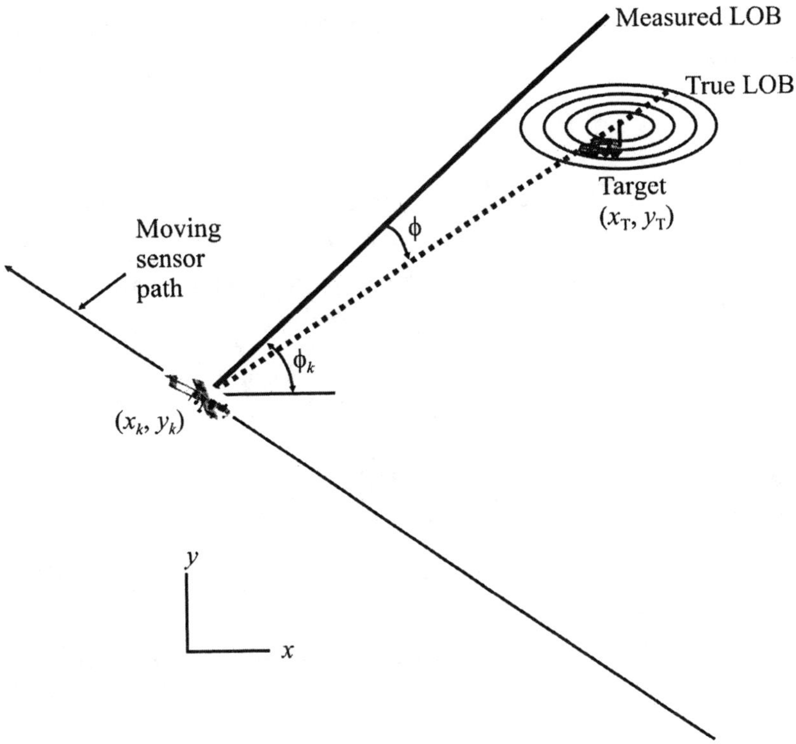

Figure 2.47 Effects of LOB bias errors.

the angle measurements are corrupted by zero mean AWGN n_k with variance σ_n^2 and bias ϕ, which has zero mean and variance σ_ϕ^2. Therefore,

$$\mathcal{E}\{n_j n_k\} = \sigma_n^2 \delta_{jk} \tag{2.319}$$

Thus, the measured bearings are characterized by

$$\phi_k = \tan^{-1}\left(\frac{\Delta_{y_k}}{\Delta_{x_k}}\right) + \phi + n_k, \qquad k = 1, 2, \ldots, N \tag{2.320}$$

where $\Delta_{y_k} = y_T - y_k$ and $\Delta_{x_k} = x_T - x_k$ and there are N observations (bearing measurements). Let

$$\boldsymbol{\theta} = [x_T \quad y_T \quad \phi]^T \tag{2.321}$$

denote the vector to be estimated. Since the bearing error bias is assumed $\sim \mathcal{N}(0, \sigma_\phi^2)$ and the noise is assumed $\sim \mathcal{N}(0, \sigma_n^2)$, then $\boldsymbol{\theta}$ has a multivariate normal distribution with mean vector $\mathbf{m}$ and positive definite covariance matrix $\boldsymbol{\Sigma}$. Let $\mathbf{h}(\mathbf{m}, \boldsymbol{\theta}) = \mathbf{0}$ be a vector of constraints. Also let

$$\mathbf{h}_m = \frac{\partial \mathbf{h}}{\partial \mathbf{m}}, \qquad \mathbf{h}_\theta = \frac{\partial \mathbf{h}}{\partial \boldsymbol{\theta}} \tag{2.322}$$

With a priori information about $\boldsymbol{\theta}$ given by the information matrix $\mathbf{J}$, the CRLB of any unbiased estimator of $\boldsymbol{\theta}$ is given by

$$\mathbf{S}_\theta = [\mathbf{h}_\theta^T (\mathbf{h}_m \boldsymbol{\Sigma} \mathbf{h}_m^T)^{-1} \mathbf{h}_\theta + \mathbf{J}]^{-1} \tag{2.323}$$

The variance of the bias, σ_ϕ^2, is assumed to be known and nonzero. Therefore,

$$\mathbf{J} = \begin{bmatrix} 0 & 0 & 0 \\ 0 & 0 & 0 \\ 0 & 0 & \dfrac{1}{\sigma_\phi^2} \end{bmatrix} \tag{2.324}$$

From (2.320),

$$\mathbf{h}_\theta = [\mathbf{G} \quad \mathbf{1}_N] \tag{2.325}$$

where

$$\mathbf{G}=\begin{bmatrix}\frac{-\Delta_{y_1}}{r_1^2} & \frac{-\Delta_{y_2}}{r_2^2} & \cdots & \frac{-\Delta_{y_N}}{r_N^2}\\ \frac{-\Delta_{x_1}}{r_1^2} & \frac{-\Delta_{x_2}}{r_2^2} & \cdots & \frac{-\Delta_{x_N}}{r_N^2}\end{bmatrix}^{\mathrm{T}} \tag{2.326}$$

with

$$r_k^2=\Delta_{x_k}^2+\Delta_{y_k}^2,\quad k=1,2,\ldots,N \tag{2.327}$$

and $\mathbf{1}_N$ is a vector of N 1's.

In this case $\mathbf{h}_m$ and $\sigma^{-2}\boldsymbol{\Sigma}$ are equal to the $N \times N$ identity matrix, since from (2.319) the noise covariance is assumed to be diagonal.

Substituting (2.325) and (2.324) into (2.323) yields

$$\mathbf{S}_\theta=\{[\mathbf{G}\quad \mathbf{1}_N]^{\mathrm{T}}(\sigma_n^2\mathbf{I}_N)^{-1}[\mathbf{G}\quad \mathbf{1}_N]+\mathrm{diag}(0,0,\sigma_\phi^{-2})\}^{-1} \tag{2.328}$$

$$=\sigma_n^2\begin{bmatrix}\mathbf{V}^{-1} & \mathbf{f}\\ \mathbf{f}^T & \left(\frac{\sigma_n}{\sigma_\phi}\right)^2+N\end{bmatrix}^{-1} \tag{2.329}$$

where

$$\mathbf{f}=\mathbf{G}^{\mathrm{T}}\mathbf{1}_N,\qquad \mathbf{V}=(\mathbf{G}^{\mathrm{T}}\mathbf{G})^{-1} \tag{2.330}$$

The CRLB is the upper left 2 × 2 block of (2.329), yielding

$$\mathbf{S}=\sigma_n^2\left[\mathbf{V}+\frac{\mathbf{V}\mathbf{f}\mathbf{f}^{\mathrm{T}}\mathbf{V}}{\left(\frac{\sigma_n}{\sigma_\phi}\right)^2+N-\mathbf{f}^{\mathrm{T}}\mathbf{V}\mathbf{f}}\right] \tag{2.331}$$

The example from [28] is useful for illustrating the effects of bias in this analysis. There is a sensor (own-ship) moving along the x-axis (y_i = 0), as illustrated in Figure 2.48. For large N, (2.330) can be expressed as

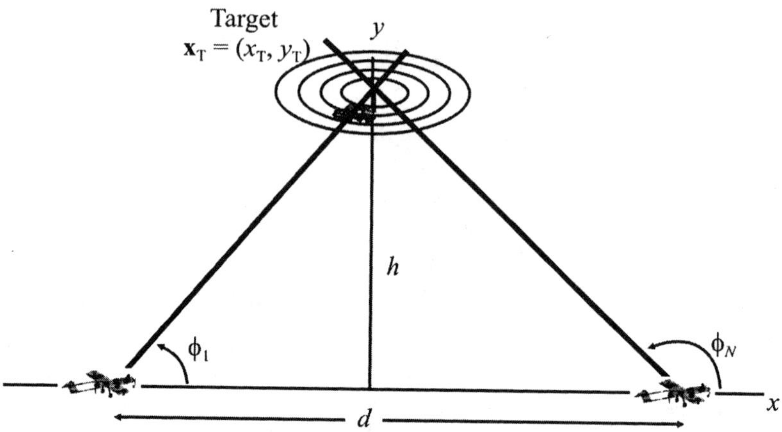

Figure 2.48 Geometry for examining the effects of bias. (*Source:* [28], © 1990 IEEE. Reprinted with permission.)

$$\mathbf{f} = \frac{N-1}{L}\begin{bmatrix} -\phi_d & \ln\dfrac{\sin\phi_N}{\sin\phi_1} \end{bmatrix}^{\mathrm{T}} \tag{2.332}$$

and

$$\mathbf{V} = \frac{2dh}{N-1}\begin{bmatrix} \dfrac{\phi_d + \sin\phi_d \cos\phi_s}{\phi_d^2 - \sin 2\phi_d} & \dfrac{\sin\phi_d \sin\phi_s}{\phi_d^2 - \sin 2\phi_d} \\ \dfrac{\sin\phi_d \sin\phi_s}{\phi_d^2 - \sin 2\phi_d} & \dfrac{\phi_d - \sin\phi_d \cos\phi_s}{\phi_d^2 - \sin 2\phi_d} \end{bmatrix} \tag{2.333}$$

where

$$\phi_d = \phi_N - \phi_1$$
$$\phi_s = \phi_1 + \phi_N$$

The CEP for this case is given by

$$\text{CEP} \approx 0.75\sqrt{\text{Tr}\mathbf{S}} \tag{2.334}$$

which is accurate to approximately 10%.

Example: For specificity, suppose that $(x_T, y_T, z_T) = (0, 50, 0)$ and the sensor trajectory is given by $x(t) = -50 + 0.15t$, $y(t) = 0$, and $z(t) = 0$. Suppose that a bearing is taken every 5 time units and that $\sigma_n = 2°$. The resulting CEP is illustrated in Figure 2.49 for the case of no bias and $\sigma_\phi = 3°$. For the case of no bias, the CEP asymptotically approaches approximately 300m while that with the bias approaches 400m.

For this example, the CEP is shown in Figure 2.50 versus the amount of bias in the bearing measurements for $T = 500$. The bias approaches

$$S_{\max} = \lim_{\sigma_\phi \to \infty} S = \sigma_n^2 \left\{ \mathbf{V} + \frac{\mathbf{Vff}^{\mathrm{T}}\mathbf{V}}{N - \mathbf{f}^{\mathrm{T}}\mathbf{Vf}} \right\} \tag{2.335}$$

■

2.11.4 Combining Noisy LOB Measurements

McCabe and Al-Samara [29] developed an approach to triangulation using a fusion methodology. The scenario considered is shown in Figure 2.51 for $N = 3$. The position estimate obtained from combining the LOBs from sensors S_1 and S_2 is given by

$$x_{12} = \frac{x_1 \tan\phi_1 - x_2 \tan\phi_2 + y_2 - y_1}{\tan\phi_1 - \tan\phi_2} \tag{2.336}$$

$$y_{12} = (x_{12} - x_1)\tan\phi_1 + y_1 \tag{2.337}$$

Likewise, the PF estimate obtained by combining the data from sensors S_2 and S_3 is given by

$$x_{23} = \frac{x_2 \tan\phi_2 - x_3 \tan\phi_3 + y_3 - y_2}{\tan\phi_2 - \tan\phi_3} \tag{2.338}$$

$$y_{23} = (x_{23} - x_2)\tan\phi_2 + y_2 \tag{2.339}$$

The covariance matrix, $\mathbf{R}_{12}$, corresponding to (2.336) and (2.337) can be approximated by

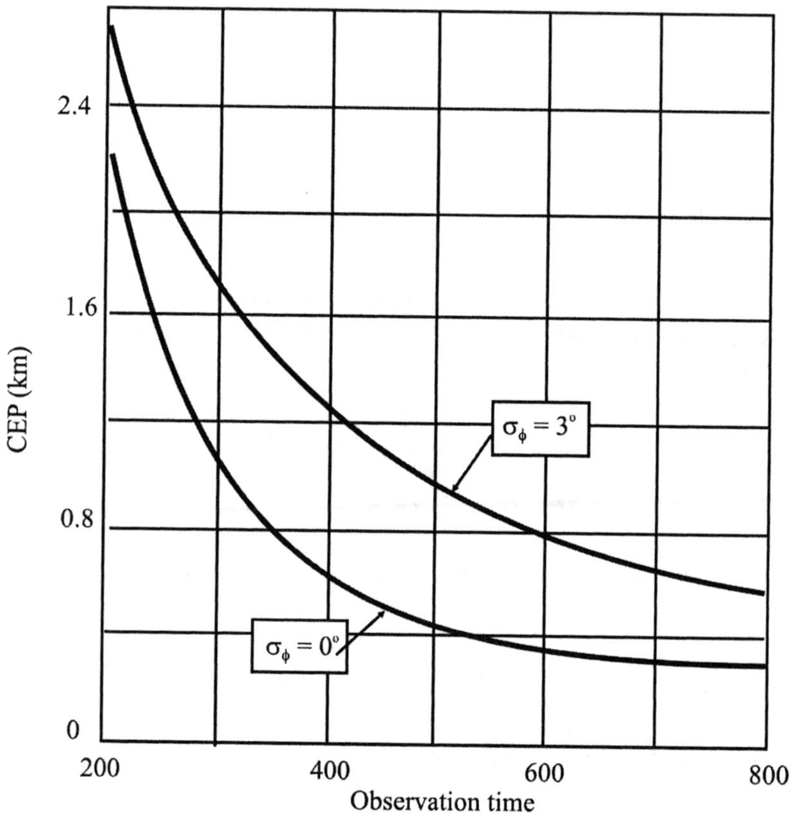

Figure 2.49 CEP of computed fixes for the bias example. (*Source:* [28], © 1990 IEEE. Reprinted with permission.)

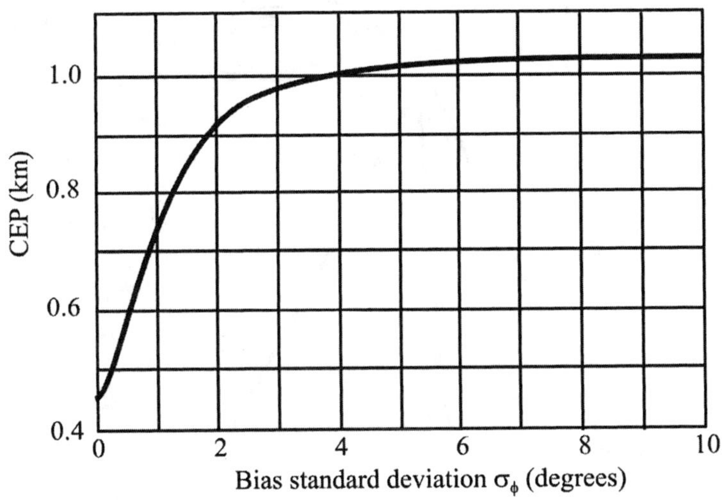

Figure 2.50 CEP versus the bias standard deviation for the example. (*Source:* [28], © 1990 IEEE. Reprinted with permission.)

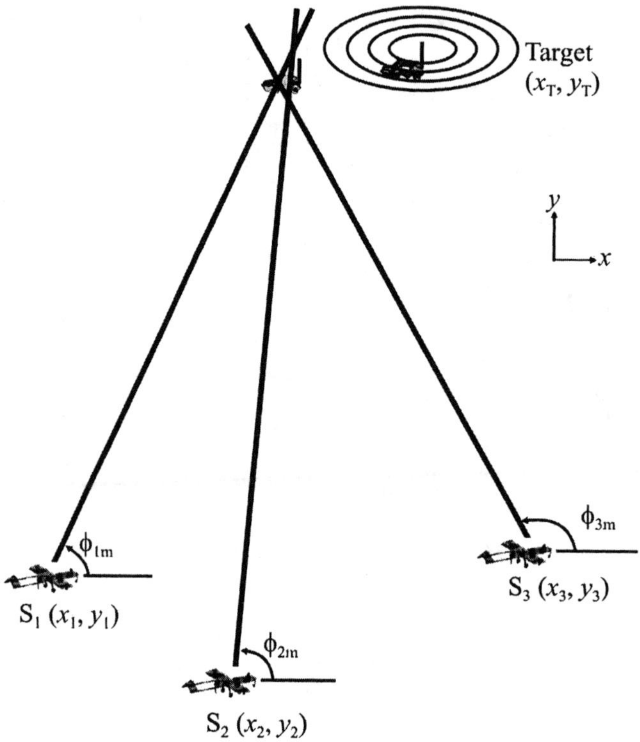

Figure 2.51 Fusion Geometry. (*Source:* [28], © 1994 IEEE. Reprinted with permission.)

$$\boldsymbol{\Delta}_{z_{12}} \triangleq \begin{bmatrix} \Delta_{x_{12}} \\ \Delta_{y_{12}} \end{bmatrix} \approx \begin{bmatrix} \dfrac{\partial x_{12}(\phi_1,\phi_2)}{\partial \phi_1} & \dfrac{\partial x_{12}(\phi_1,\phi_2)}{\partial \phi_2} \\ \dfrac{\partial y_{12}(\phi_1,\phi_2)}{\partial \phi_1} & \dfrac{\partial y_{12}(\phi_1,\phi_2)}{\partial \phi_2} \end{bmatrix} \begin{bmatrix} \Delta_{\theta_1} \\ \Delta_{\theta_2} \end{bmatrix} \tag{2.340}$$

$$\triangleq \mathbf{A}_{12} \begin{bmatrix} \Delta_{\theta_1} \\ \Delta_{\theta_2} \end{bmatrix} \tag{2.341}$$

and

$$\mathbf{R}_{12} \triangleq \mathcal{E}\{\boldsymbol{\Delta}_{z_{12}} \boldsymbol{\Delta}_{z_{12}}^{\mathrm{T}}\} = \mathbf{A}_{12} \begin{bmatrix} \sigma_{\phi_1}^2 & 0 \\ 0 & \sigma_{\phi_2}^2 \end{bmatrix} \mathbf{A}_{12}^{\mathrm{T}} \tag{2.342}$$

Likewise,

$$\mathbf{R}_{23} \triangleq \mathcal{E}\{\boldsymbol{\Delta}_{z_{23}} \boldsymbol{\Delta}_{z_{23}}^{\mathrm{T}}\} = \mathbf{A}_{23} \begin{bmatrix} \sigma_{\phi_2}^2 & 0 \\ 0 & \sigma_{\phi_3}^2 \end{bmatrix} \mathbf{A}_{23}^{\mathrm{T}} \tag{2.343}$$

The cross-covariance matrix, $\mathbf{R}_{123}$, corresponding to these measurements is

$$\mathbf{R}_{123} \triangleq \mathcal{E}\{\boldsymbol{\Delta}_{z_{12}} \boldsymbol{\Delta}_{z_{23}}^{\mathrm{T}}\} = \mathbf{A}_{12} \begin{bmatrix} 0 & 0 \\ \sigma_{\phi_2}^2 & 0 \end{bmatrix} \mathbf{A}_{23} \tag{2.344}$$

The combining of these measurements is done with a fusion algorithm from [30]. That algorithm forms the linear combination

$$\hat{\mathbf{X}} = \sum_{i=1}^{M} \mathbf{A}_i \hat{\mathbf{X}}_i \tag{2.345}$$

For $N = 2$, as considered here,

$$\mathbf{A}_1 = \mathbf{R}_{21}'[\mathbf{R}']^{-1} \tag{2.346}$$

and

$$\mathbf{A}_2 = [\mathbf{I} - \mathbf{A}_1] \tag{2.347}$$

where

$$\mathbf{R}'_{21} = \mathbf{R}_{23} - \mathbf{R}^{\mathrm{T}}_{123}$$
$$\mathbf{R}' = \mathbf{R}_{12} - \mathbf{R}_{123} - \mathbf{R}'_{123}$$

The resulting fused covariance matrix is given by

$$\mathbf{R}_0 = \mathbf{R}_{23} - (\mathbf{R}_{23} - \mathbf{R}^{\mathrm{T}}_{123})(\mathbf{R}_{12} - \mathbf{R}_{123} - \mathbf{R}^{\mathrm{T}}_{123} + \mathbf{R}_{23})^{-1}(\mathbf{R}^{\mathrm{T}}_{23} - \mathbf{R}_{123}) \tag{2.348}$$

Although presented in two dimensions, this algorithm extends to three dimensions in the obvious way. Also, only three sensors were illustrated yielding $N = 2$ sets of LOB measurements. This too extends in the natural way to $N > 2$.

Example: Consider the case where the pdfs for the LOBs at the sensors are given as in Figure 2.52. Three sensors are located at S_1: (2, 3); S_2: (5, 4); and S_3: (9, 3). The pdf parameters are $\delta_1 = 5^\circ$, $\delta_2 = 4^\circ$, and $\delta_3 = 6^\circ$. The fused x-data errors are shown in Figure 2.53(a) and the fused y-data errors are shown in Figure 2.53(b). ■

2.11.5 Effects of Navigation Error

Another source of error in triangulation estimation of the PF is errors associated with knowing the locations of the sensors. Manolakis and Cox [31] presented an analysis of sensor position error effects on the estimation of three-dimensional target PFs, although their analysis was actually based on calculating the effects of range difference estimation using *distance measuring equipment* (DME) to estimate the location of cooperating aircraft. This section follows [31] closely.

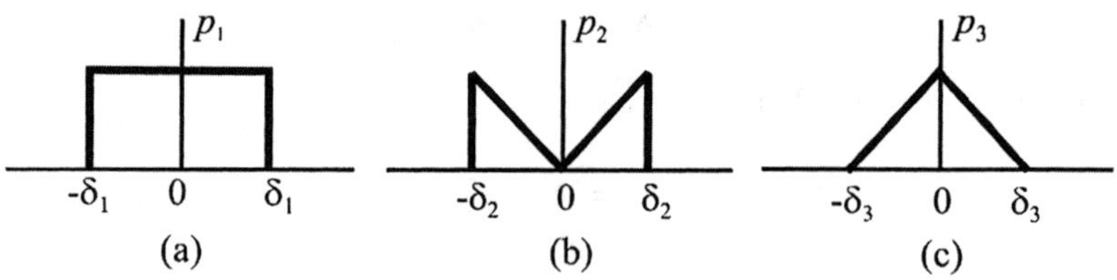

Figure 2.52 Pdfs for the three sensors in the example. (*Source:* [29], © 1994 IEEE. Reprinted with permission.)

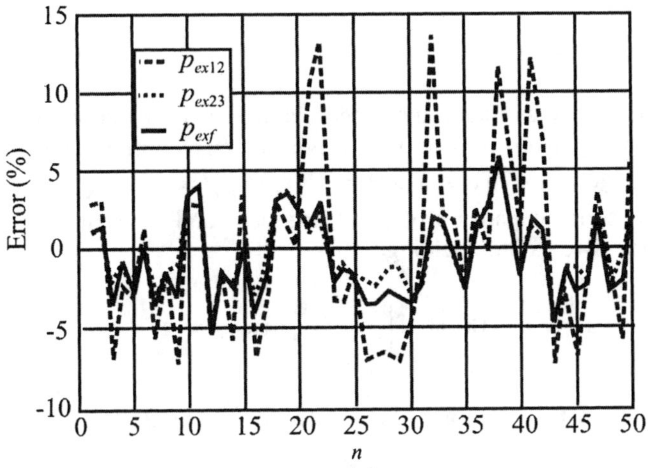

(a)

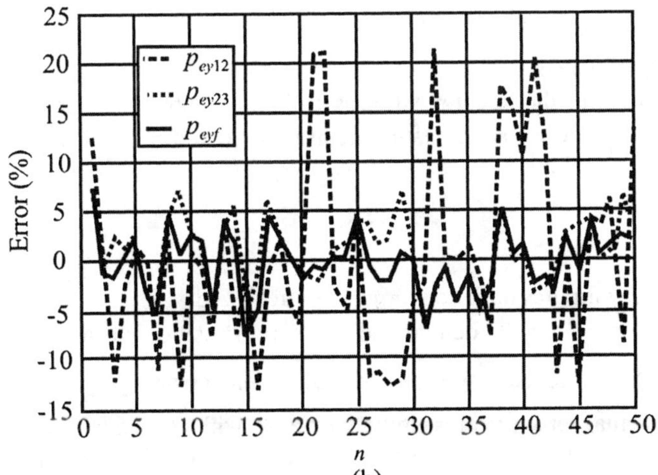

(b)

Figure 2.53 Results of the simulation for the example. Chart (a) illustrates the errors for the x-parameters, while chart (b) is for the y-parameters. (*Source:* [29], © 1994 IEEE. Reprinted with permission.)

The notation used here is as follows:

$\mathbf{x}_i = [x_i \ y_i \ z_i]^{\mathrm{T}}$ is the vector of coordinates for sensor i
$\mathbf{x}_{\mathrm{T}} = [x_{\mathrm{T}} \ y_{\mathrm{T}} \ z_{\mathrm{T}}]^{\mathrm{T}}$ is the vector of target coordinates
$f_i(\mathbf{x}_{\mathrm{T}})$ is the system measurement equation for sensor i given by

$$f_i(\mathbf{x}_\text{T}) = 0 \tag{2.349}$$

For simplicity it is assumed that there is one master sensor where the PF calculations are performed and two slave stations configured similar to that illustrated in Figure 2.51. Thus, $\mathbf{f}(x_\text{T}) = [f_1(x_\text{T}) \quad f_2(x_\text{T}) \quad f_3(x_\text{T})]^\text{T}$.

A Taylor series expansion of (2.349) is used and the terms higher than the linear ones are disregarded, yielding

$$\mathbf{f}(\mathbf{x}_\text{T}) \approx \mathbf{f}\Big|_{\mathbf{x}_{T_0}} + \mathbf{G}\Big|_{\mathbf{x}_{T_0}} (\mathbf{x}_\text{T} - \mathbf{x}_{\text{T}_0}) \tag{2.350}$$

where the notation $\mathbf{G}\Big|_{\mathbf{x}_{T_0}}$ denotes $\mathbf{G}$ evaluated at $\mathbf{x}_{\text{T}_0}$. In (2.350),

$$\mathbf{G} = \frac{\partial \mathbf{f}}{\partial \mathbf{x}_\text{T}} \tag{2.351}$$

is the Jacobian matrix.

Calculation of the target position is an iterative process, with the initial guess denoted by $\mathbf{x}_{\text{T}_0}$. New position estimates are provided by (2.349) and (2.350) as

$$\mathbf{x}_\text{T} \approx \mathbf{x}_{\text{T}_0} - \mathbf{G}^{-1}\mathbf{f}_{|\mathbf{x}_{\text{T}_0}} \tag{2.352}$$

and iteration continues until convergence is achieved.

Let $[x_{s_j} \quad y_{s_j} \quad x_{s_j}]^\text{T}$ denote the measured position vector of sensor j and $[\delta_{x_j} \quad \delta_{y_j} \quad \delta_{z_j}]^\text{T}$ the corresponding measurement errors for j = 1, 2, 3. Let $\mathbf{p}_x$ denote the actual sensor vector along the x-axis, with corresponding definitions for measured sensor vectors along the y-axis and z-axis. Thus, using this notation, $\mathbf{p}_x = [x_1 \quad x_2 \quad x_3]^\text{T}$, $\mathbf{p}_y = [y_1 \quad y_2 \quad y_3]^\text{T}$, and $\mathbf{p}_z = [z_1 \quad z_2 \quad z_3]^\text{T}$. Let $\mathbf{p}_{sx}$ denote the vector of measured sensor x-coordinates so that $\mathbf{p}_{sx} = [x_{s_1} \quad x_{s_2} \quad x_{s_3}]^\text{T}$. Likewise, $\mathbf{p}_{sy} = [y_{s_1} \quad y_{s_2} \quad y_{s_3}]^\text{T}$ and $\mathbf{p}_{sz} = [z_{s_1} \quad z_{s_2} \quad z_{s_3}]^\text{T}$. Last, let $\delta\mathbf{p}_x$ denote the error vector containing the errors in measuring the coordinates, viz, $\delta\mathbf{p}_{sx} = [\delta_{x_1} \quad \delta_{x_2} \quad \delta_{x_3}]^\text{T}$, as well as $\delta\mathbf{p}_{sy} = [\delta_{y_1} \quad \delta_{y_2} \quad \delta_{y_3}]^\text{T}$, and $\delta\mathbf{p}_{sz} = [\delta_{z_1} \quad \delta_{z_2} \quad \delta_{z_3}]^\text{T}$. Obviously,

$$\mathbf{p}_{si} = \mathbf{p}_i + \delta\mathbf{p}_i \qquad i \in \{x, y, z\} \tag{2.353}$$

Let $\mathbf{p} = [\mathbf{p}_x \quad \mathbf{p}_y \quad \mathbf{p}_z]^{\mathrm{T}}$ denote the vector consisting of all of the coordinates. That is, if $\mathbf{p}_x$ is 3 × 1, then $\mathbf{p}$ is 9 × 1. Likewise for $\mathbf{p}_s$ and $\delta\mathbf{p}$.

The errors in estimating the PF of the target due to sensor positioning errors are given by $\boldsymbol{\varepsilon} = [\varepsilon_x \quad \varepsilon_y \quad \varepsilon_z]^{\mathrm{T}}$ and

$$\boldsymbol{\varepsilon} = \mathbf{x}'_{\mathbf{x}_\mathrm{T}} - \mathbf{x}_\mathrm{T} \tag{2.354}$$

where

$$\mathbf{x}'_{\mathbf{x}_\mathrm{T}} = \mathbf{x}_{\mathrm{T}_0} - \mathbf{G}^{-1}\mathbf{f}\Big|_{(\mathbf{x}_{\mathrm{T}_0}, \mathbf{p}_s)} \tag{2.355}$$

is the estimate of the PF when $\mathbf{G}^{-1}$ and $\mathbf{f}$ are evaluated at $(\mathbf{x}_{\mathrm{T}_0}, \mathbf{p}_s)$.

Therefore, errors in sensor positioning can and should be included in (2.349) yielding

$$\mathbf{f}(\mathbf{x}_\mathrm{T}, \mathbf{p}) \approx \mathbf{f} + \mathbf{G}(\mathbf{x}_\mathrm{T} - \mathbf{x}_{\mathrm{T}_0})\Big|_{(\mathbf{x}_{\mathrm{T}_0}, \mathbf{p}_s)} + \left(\frac{\partial \mathbf{f}}{\partial \mathbf{p}}\right)^{\mathrm{T}} \delta\mathbf{p}\Bigg|_{(\mathbf{x}_{\mathrm{T}_0}, \mathbf{p}_s)} \tag{2.356}$$

so that the new value of $\mathbf{x}_\mathrm{T}$ is given by

$$\mathbf{x}_\mathrm{T} = \mathbf{x}_{\mathrm{T}_0} - \mathbf{G}^{-1}\mathbf{f}\Big|_{(\mathbf{x}_{\mathrm{T}_0}, \mathbf{p}_s)} - \mathbf{G}^{-1}\left(\frac{\partial \mathbf{f}}{\partial \mathbf{p}}\right)^{\mathrm{T}} \delta\mathbf{p}\Bigg|_{(\mathbf{x}_{\mathrm{T}_0}, \mathbf{p}_s)} \tag{2.357}$$

or

$$\mathbf{x}_\mathrm{T} = \mathbf{x}'_{\mathrm{T}_0} - \mathbf{G}^{-1}\left(\frac{\partial \mathbf{f}}{\partial \mathbf{p}}\right)^{\mathrm{T}} \delta\mathbf{p}\Bigg|_{(\mathbf{x}_{\mathrm{T}_0}, \mathbf{p}_s)} \tag{2.358}$$

yielding

$$\boldsymbol{\varepsilon} = -\mathbf{G}^{-1}\left(\frac{\partial \mathbf{f}}{\partial \mathbf{p}}\right)^{\mathrm{T}} \delta\mathbf{p}\Bigg|_{(\mathbf{x}_{\mathrm{T}_0},\mathbf{p}_s)} \tag{2.359}$$

Consider the case of errors in sensor positioning on PF estimation using triangulation with generalized bearings as discussed in Section 2.6. The geometry is shown in Figure 2.29, repeated here as Figure 2.54. The system equations are given by

$$f_i(\mathbf{x}_{\mathrm{T}},\mathbf{p}_i) = 0 = \tan^{-1}\frac{\sqrt{(x_{\mathrm{T}}-x_i)^2+(z_{\mathrm{T}}-z_i)^2}}{y_{\mathrm{T}}-y_i} - \alpha_i \tag{2.360}$$

where α_i is the measured generalized bearing from $\mathbf{x}_i$ to $\mathbf{x}_{\mathrm{T}}$. Now,

$$f_i(x_{\mathrm{T}_0}+\delta x, y_{\mathrm{T}_0}+\delta y, z_{\mathrm{T}_0}+\delta z, \mathbf{p}_i)$$
$$= f_i(x_{\mathrm{T}_0}, y_{\mathrm{T}_0}, z_{\mathrm{T}_0}, \mathbf{p}_i) + \left(\delta x\frac{\partial}{\partial x_{\mathrm{T}}} + \delta y\frac{\partial}{\partial y_{\mathrm{T}}} + \delta z\frac{\partial}{\partial z_{\mathrm{T}}}\right) f_i(x_{\mathrm{T}}, y_{\mathrm{T}}, z_{\mathrm{T}}, \mathbf{p}_i) \tag{2.361}$$

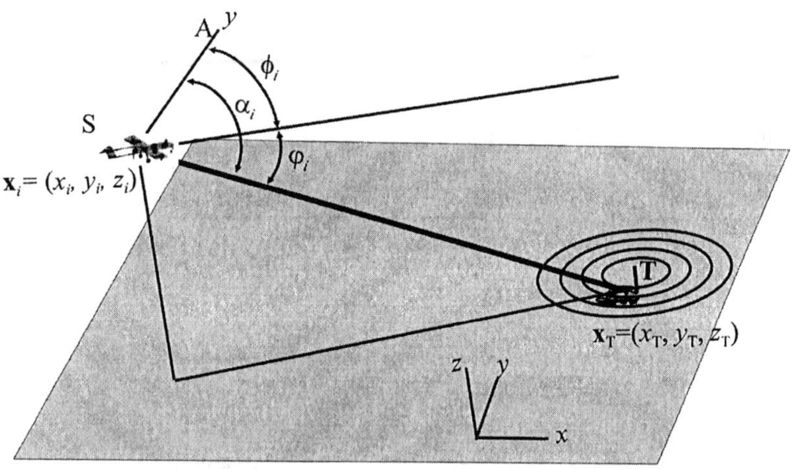

Figure 2.54 Generalized bearing, denoted by α_i.

In (2.361),

$$f_i(x_{T_0}, y_{T_0}, z_{T_0}, \mathbf{p}_i) = \tan^{-1}\frac{\sqrt{(x_{T_0}-x_i)^2+(z_{T_0}-z_i)^2}}{y_{T_0}-y_i} - \alpha_i \tag{2.362}$$

$$\frac{\partial f_i}{\partial x_T} = \frac{(x_T-x_i)(y_T-y_i)}{(x_T-x_i)^2+(y_T-y_i)^2+(z_T-z_i)^2}\frac{1}{\sqrt{(x_T-x_i)^2+(z_T-z_i)^2}} \tag{2.363}$$

$$\frac{\partial f_i}{\partial y_T} = -\frac{\sqrt{(x_T-x_i)^2+(z_T-z_i)^2}}{\sqrt{(x_T-x_i)^2+(y_T-y_i)^2+(z_T-z_i)^2}} \tag{2.364}$$

$$\frac{\partial f_i}{\partial z_T} = \frac{(z_T-z_i)(y_T-y_i)}{(x_T-x_i)^2+(y_T-y_i)^2+(z_T-z_i)^2}\frac{1}{\sqrt{(x_T-x_i)^2+(z_T-z_i)^2}} \tag{2.365}$$

which are all evaluated at $x_T = x_{T_0}, y_T = y_{T_0}$, and $z_T = z_{T_0}$.

The Jacobian matrix in this case is

$$\mathbf{G}\big|_{\mathbf{x}_{T_0}} = \frac{\partial \mathbf{f}}{\partial \mathbf{x}_T}\bigg|_{\mathbf{x}_{T_0}} \triangleq \begin{bmatrix} \left.\frac{\partial f_1}{\partial x_T}\right|_{\substack{\mathbf{x}_T=\mathbf{x}_{T_0}\\ i=1}} & \left.\frac{\partial f_2}{\partial x_T}\right|_{\substack{\mathbf{x}_T=\mathbf{x}_{T_0}\\ i=2}} & \left.\frac{\partial f_3}{\partial x_T}\right|_{\substack{\mathbf{x}_T=\mathbf{x}_{T_0}\\ i=3}} \\ \left.\frac{\partial f_1}{\partial y_T}\right|_{\substack{\mathbf{x}_T=\mathbf{x}_{T_0}\\ i=1}} & \left.\frac{\partial f_2}{\partial y_T}\right|_{\substack{\mathbf{x}_T=\mathbf{x}_{T_0}\\ i=2}} & \left.\frac{\partial f_3}{\partial y_T}\right|_{\substack{\mathbf{x}_T=\mathbf{x}_{T_0}\\ i=3}} \\ \left.\frac{\partial f_1}{\partial z_T}\right|_{\substack{\mathbf{x}_T=\mathbf{x}_{T_0}\\ i=1}} & \left.\frac{\partial f_2}{\partial z_T}\right|_{\substack{\mathbf{x}_T=\mathbf{x}_{T_0}\\ i=2}} & \left.\frac{\partial f_3}{\partial z_T}\right|_{\substack{\mathbf{x}_T=\mathbf{x}_{T_0}\\ i=3}} \end{bmatrix} \tag{2.366}$$

with $\partial f_i / \partial x_T\big|_{\substack{\mathbf{x}_T=\mathbf{x}_{T_0}\\ i=k}}$ given by (2.363) with $i = k$, $\partial f_i / \partial y_T\big|_{\substack{\mathbf{x}_T=\mathbf{x}_{T_0}\\ i=k}}$ given by (2.364) with $i = k$, and $\partial f_i / \partial z_T\big|_{\substack{\mathbf{x}_T=\mathbf{x}_{T_0}\\ i=k}}$ given by (2.365) with $i = k$.

The derivative of **f** with respect to **p** is given by

$$\frac{\partial \mathbf{f}}{\partial \mathbf{p}} \triangleq \begin{bmatrix} \frac{\partial \mathbf{f}_1}{\partial \mathbf{p}_x} & \frac{\partial \mathbf{f}_2}{\partial \mathbf{p}_x} & \frac{\partial \mathbf{f}_3}{\partial \mathbf{p}_x} \\ \frac{\partial \mathbf{f}_1}{\partial \mathbf{p}_y} & \frac{\partial \mathbf{f}_2}{\partial \mathbf{p}_y} & \frac{\partial \mathbf{f}_3}{\partial \mathbf{p}_y} \\ \frac{\partial \mathbf{f}_1}{\partial \mathbf{p}_z} & \frac{\partial \mathbf{f}_2}{\partial \mathbf{p}_z} & \frac{\partial \mathbf{f}_3}{\partial \mathbf{p}_z} \end{bmatrix} \tag{2.367}$$

$$= \begin{bmatrix} \frac{\partial f_1}{\partial x_1} & & & \frac{\partial f_1}{\partial y_1} & & & \frac{\partial f_1}{\partial z_1} & & \\ & \frac{\partial f_2}{\partial x_2} & & & \frac{f_2}{\partial y_2} & & & \frac{\partial f_2}{\partial z_2} & \\ & & \frac{\partial f_3}{\partial x_3} & & & \frac{\partial f_3}{\partial x_3} & & & \frac{f_3}{\partial z_3} \end{bmatrix}^{\mathrm{T}} \tag{2.368}$$

with

$$\frac{\partial f_j}{\partial x_i} = \begin{cases} -\dfrac{(x_{\mathrm{T}} - x_i)(y_{\mathrm{T}} - y_i)}{(x_{\mathrm{T}} - x_i)^2 + (y_{\mathrm{T}} - y_i)^2 + (z_{\mathrm{T}} - z_i)^2} \times \dfrac{1}{\sqrt{(x_{\mathrm{T}} - x_i)^2 + (z_{\mathrm{T}} - z_i)^2}}, & i = j \\ 0, & \text{otherwise} \end{cases} \tag{2.369}$$

$$\frac{\partial f_j}{\partial y_i} = \begin{cases} \dfrac{\sqrt{(x_{\mathrm{T}} - x_i)^2 + (z_{\mathrm{T}} - z_i)^2}}{(x_{\mathrm{T}} - x_i)^2 + (y_{\mathrm{T}} - y_i)^2 + (z_{\mathrm{T}} - z_i)^2}, & i = j \\ 0, & \text{otherwise} \end{cases} \tag{2.370}$$

$$\frac{\partial f_j}{\partial z_i} = \begin{cases} -\dfrac{(y_{\mathrm{T}} - y_i)(z_{\mathrm{T}} - z_i)}{(x_{\mathrm{T}} - x_i)^2 + (y_{\mathrm{T}} - y_i)^2 + (z_{\mathrm{T}} - z_i)^2} \times \dfrac{1}{\sqrt{(x_{\mathrm{T}} - x_i)^2 + (z_{\mathrm{T}} - z_i)^2}}, & i = j \\ 0, & \text{otherwise} \end{cases} \tag{2.371}$$

These derivatives are all evaluated at $\mathbf{x}_T = \mathbf{x}_{T_0}$. This process is iterated as long as necessary after updating the target location estimate with (2.358).

2.12 Concluding Remarks

The notion of geolocating emitting targets by triangulation was introduced in this chapter. Triangulation requires processing LOBs and determining the most likely target location based on where the LOBs intersect. When there is no noise or measurement error, this intersection is at a single point and the optimal PF is at that point. Such is rarely the case, however, and noise and measurement error must be taken into account.

Triangulation is based on relatively simple trigonometric principles. Determination of the optimum PF, taking noise into consideration, complicates the calculations. Least-squared estimation techniques are quite common for calculating optimum locations, in a least-squared sense. Least-squared means that the error that is the square of the difference between the actual parameter being estimated and the estimate is minimized.

A novel PF calculation method based on the discrete probability density using the Tikhonov was presented. This method deals with inaccurate data and tends to minimize the effects of highly inaccurate LOBs.

The Kalman filter realizations for PF determination are sequential methods for PF determination. They typically assume that the data is arriving at the sensor in a serial fashion, and the estimate of the PF is updated based on newly arriving data. The PF calculation problem is normally nonlinear in nature, and the standard Kalman filter does not apply. In that case, the extended Kalman filter can be used which linearizes the problem with a Taylor series expansion of the nonlinear equations.

The maximum likelihood methods attempt to find the optimal PF based on finding the location with maximizing probability. The a priori probabilities are assumed or measured, and the data is used to maximize those pdfs. Stansfield's method was perhaps the first of these techniques and is widely used for PF calculation. It is a biased technique, however, in that as the amount of data is increased, the covariance does not tend to zero.

There are many sources of error in triangulation. Some are due to the aforementioned noise and measurement error. Others are due to the geometry of the problem and can only be addressed by changing the geometry. GDOP is the primary cause for these types of errors.

A summary comparison of the methods introduced in this chapter is provided in Table 2.5. It should be noted that for most of these methods, special conditions may exist that lower the computational complexity. Indeed, it is the search for

Table 2.5 Comparison of Triangulation Methods Discussed

Method	Biased	Covariance	Iterative	Complexity	Remarks
Least-squares — Brown	No	(2.59)	No	$\mathcal{O}(MN^2 \log N)$ [32]	Can be iterated and/or recursive
Least-squares — Hemispheric	No	(2.83)	Yes	$\mathcal{O}(MN^2 \log N)$ [32]	Can be iterated and/or recursive
Least-squares — Pages-Zamora	No	(2.103)	No	$\mathcal{O}(MN^2 \log N)$ [32]	Can be iterated and/or recursive
TLS	No	N/A	No	$\mathcal{O}(N^3)$ [33]	$\mathcal{O}(N)$ or $\mathcal{O}(N^2)$ using minor component analysis
MMSE	No	(2.128)	No	$\mathcal{O}(MN^2 \log N)$	
LMMSE	No	(2.153)	No	$\mathcal{O}(MN^2 \log N)$	
Kalman Filter	N/A	(2.215)	Yes	$\mathcal{O}(MN^2 \log N)$ [34]	
DPD	No	N/A	No	$\mathcal{O}(NL^2)$	Grid search required L is the number of grid cells in each dimension
Generalized Bearings	N/A	N/A	Yes	$\mathcal{O}(MN^2 \log N)$	
MLE	No	(2.255)	Yes	N/A	
Stansfield	(2.292)	(2.293)	No	N/A	Widely used
Multiple Sample Correlation	N/A	N/A	Yes	N/A	Must know or estimate limits of LOB accuracy

All complexity estimates are per iteration when applicable.
M: Number of sensors.
N: Number of data points.
N/A: Not available.
In all cases zero mean AWGN is assumed.

these special conditions that is the focus of much of the modern research in PF estimation.

Appendix 2A

Least-Squared Error Estimation Program Listing

This appendix contains a source program listing in MathCad® for the least-squared algorithm discussed in Section 2.2. This algorithm presented in Section 2.2 is intended to be a multipass, iterative procedure where the output of each pass is used in the subsequent pass to refine the PF calculation.

Initialize variables

$\psi_T := 15$ $\quad\quad\quad i := 0..20$

$\lambda_T := 15$

$\lambda hat_0 := 13$

$\psi hat_0 := 13$

$\lambda_S(i) := i$

$\psi_S(t) := 0$

$$sigma := 2 \cdot \frac{\pi}{180}$$

Calculate the bearing errors according to the normal distribution.

BearingError := rnorm(21, 0, sigma)

Calculate the bearings to which the random errors will be added.

$$\phi_i := acos\left[\frac{\left(\psi_T - \psi_S(i)\right)}{\sqrt{\left(\psi_T - \psi_S(i)\right)^2 + \left(\lambda_T - \lambda_S(i)\right)^2}}\right]$$

$\phi =$

	0
0	0.785
1	0.751
2	0.714
3	0.675
4	0.633
5	0.588
6	0.54
7	0.49
8	0.437
9	0.381
10	0.322
11	0.261
12	0.197
13	0.133
14	0.067
15	0

ϕtilde is the vector of measured bearings.

$\phi\text{tilde} := \phi + \text{BearingError}$

Iteration

Initialize nhat and mhat for this pass. With multiple iterations, λhat.0 and ϕhat.0 are updated with the results from the previous pass.

$\text{nhat}_i := \lambda\text{hat}_0 - \lambda_s(i)$

Calculate the estimated bearing vector.

$\text{mhat}_i := \psi\text{hat}_0 - \psi_s(i)$

$$\phi hat_i := acos\left[\frac{mhat_i}{\sqrt{(mhat_i)^2 + (nhat_i)^2}}\right]$$

Calculate vector Y.

$$Y := \phi tilde - \phi hat$$

Determine the elements of two columns of X.

$$X_{t1_i} := \frac{nhat_i}{(mhat_i)^2 + (nhat_i)^2}$$

$$X_{t2_i} := \frac{mhat_i}{(mhat_i)^2 + (nhat_i)^2}$$

Convert these to vectors.

$$X_{i,0} := X_{t1_i}$$

$$X_{i,1} := X_{t2_i}$$

Calcuate the error vector β.

$$\beta hat := \left(X^T \cdot X\right)^{-1} X^T \cdot Y$$

Calculate the estimated PF for this pass.

$$P_{T_0} := \psi hat_0 - \beta hat_0$$

$$P_{T_1} := \lambda hat_0 - \beta hat_1$$

$$P_T = \begin{pmatrix} 10.256 \\ 13.415 \end{pmatrix}$$

$\psi hat_0 := P_{T_0}$

$\lambda hat_0 := P_{T_1}$

Iterate

Appendix 2B

Generalized Bearing Program Listing

This appendix contains a source program listing in MathCad® for the generalized bearing algorithm discussed in Section 2.7. This algorithm is intended to be a multipass, iterative procedure where the output of each pass is used in the subsequent pass to refine the PF calculation.

Initialize variables

$m := 0$

$y_T := 0$

$i := 0..20$

$x_T := 10$

$\rho_{\theta\varepsilon} := 0$

$$\sigma_\theta := 2 \cdot \frac{\pi}{180}$$

$$\sigma_\varepsilon := 2 \cdot \frac{\pi}{180}$$

$z_T := 0.5$

Distances are in kilometers

Initial values of estimated location

$xhat_0 := 12$

$yhat_0 := 5$

$zhat_0 := 0.5$

$$xhat_{v0} := \begin{pmatrix} xhat_0 \\ yhat_0 \\ zhat_0 \end{pmatrix}$$

Equations of sensor motion. An LOB is measured every unit of distance.

$x_s(i) := i$

$y_s(i) := 0$

$z_s(i) := 3$

Calculate the bearing errors according to the normal distribution.

$\text{BearingError} := \text{rnorm}(21, 0, \sigma_\theta)$

Calculate the generalized bearings to which the random errors will be added.

$$\alpha_i := \text{atan}\left[\frac{\sqrt{(x_T - x_s(i))^2 + (z_T - z_s(i))^2}}{y_T - y_s(i)}\right]$$

The vector of observed bearings

$\alpha_{obs} := \alpha + \text{BearingError}$

Iteration

$m := m + 1$

Calculate the components of Gm.

$$r_m(i) := \sqrt{(xhat_0 - x_s(i))^2 + (yhat_0 - y_s(i))^2 + (zhat_0 - z_s(i))^2}$$

$$s_m(i) := \sqrt{\left(\frac{r_m(i)}{yhat_0 - y_s(i)}\right)^2 - 1}$$

$$g_{m_{i,0}} := \frac{-(xhat_0 - x_s(i))}{(r_m(i))^2 \cdot s_m(i)} \qquad g_{m_{i,1}} := \frac{(yhat_0 - y_s(i))s_m(i)}{r_m(i)}$$

$$G_m := \begin{pmatrix} g_{m_{0,0}} & g_{m_{0,1}} & g_{m_{0,2}} \\ g_{m_{1,0}} & g_{m_{1,1}} & g_{m_{1,2}} \\ g_{m_{2,0}} & g_{m_{2,1}} & g_{m_{2,2}} \\ g_{m_{3,0}} & g_{m_{3,1}} & g_{m_{3,2}} \\ g_{m_{4,0}} & g_{m_{4,1}} & g_{m_{4,2}} \\ g_{m_{5,0}} & g_{m_{5,1}} & g_{m_{5,2}} \\ g_{m_{6,0}} & g_{m_{6,1}} & g_{m_{6,2}} \\ g_{m_{7,0}} & g_{m_{7,1}} & g_{m_{7,2}} \\ g_{m_{8,0}} & g_{m_{8,1}} & g_{m_{8,2}} \\ g_{m_{9,0}} & g_{m_{9,1}} & g_{m_{9,2}} \\ g_{m_{10,0}} & g_{m_{10,1}} & g_{m_{10,2}} \\ g_{m_{11,0}} & g_{m_{11,1}} & g_{m_{11,2}} \\ g_{m_{12,0}} & g_{m_{12,1}} & g_{m_{12,2}} \\ g_{m_{13,0}} & g_{m_{13,1}} & g_{m_{13,2}} \\ g_{m_{14,0}} & g_{m_{14,1}} & g_{m_{14,2}} \\ g_{m_{15,0}} & g_{m_{15,1}} & g_{m_{15,2}} \\ g_{m_{16,0}} & g_{m_{16,1}} & g_{m_{16,2}} \\ g_{m_{17,0}} & g_{m_{17,1}} & g_{m_{17,2}} \\ g_{m_{18,0}} & g_{m_{18,1}} & g_{m_{18,2}} \\ g_{m_{19,0}} & g_{m_{19,1}} & g_{m_{19,2}} \\ g_{m_{20,0}} & g_{m_{20,1}} & g_{m_{20,2}} \end{pmatrix}$$

Calculate the correlation coefficients.

$$a_m(i) := \left[\frac{-1}{s_m(i)} \cdot \frac{\left(xhat_0 - x_s(i)\right)}{yhat_0 - y_s(i)}\right]^2$$

$$b_m(i) := \left[s_m(i) \cdot \sqrt{\left(\frac{r_m(i)}{zhat_0 - z_s(i)}\right)^2 - 1}\right]^{-2}$$

$$c_m(i) := \sqrt{a_m(i) \cdot b_m(i)}$$

$$\sigma_m(i) := \sqrt{a_m(i) \cdot \left(\sigma_\theta\right)^2 + b_m(i) \cdot \left(\sigma_\varepsilon\right)^2 + c_m(i) \cdot \rho_{\theta\varepsilon} \cdot \sigma_\theta \cdot \sigma_\varepsilon}$$

Convert σ to a vector.

$$h_i := \sigma_m(i)$$

$$C := \mathrm{diag}(h)$$

$$\alpha hat_i := \mathrm{atan}\left[\sqrt{\frac{\left(xhat_0 - x_s(i)\right)^2 + \left(zhat_0 - z_s(i)\right)^2}{yhat_0 - y_s(i)}}\right]$$

$$w_m := \alpha_{obs} + G_m \cdot xhat_{v0} - \alpha hat$$

Calculate the updated location estimate.

$$xhat_1 := \left(G_m^T C^{-1} G_m\right)^{-1} G_m^T C^{-1} w_m$$

$$xhat_1 = \begin{pmatrix} 13.207 \\ 4.254 \\ 2.112 \end{pmatrix}$$

$$xhat_0 := xhat_{1_0}$$

$$yhat_0 := xhat_{1_1}$$

$$zhat_0 := xhat_{1_2}$$

The above procedure is iterated until the difference in location estimate is small.

References

[1] Jenkins, H. H., *Small Aperture Radio Direction Finding,* Norwood, MA: Artech House, 1991, p. 185.

[2] Cadzow, J. A., "Signal Processing via Least-Squared Error Modeling," *IEEE ASSP Magazine,* October 1990, pp. 12–31.

[3] Mendel, J. M., *Lessons in Estimation Theory for Signal Processing, Communications, and Control,* Englewood Cliffs, NJ: Prentice-Hall, 1995, pp. 115–117.

[4] Golub, G. H., and C. F. Van Loan, "An Analysis of the Total Least-Squared Problem," *SIAM Journal on Numerical Analysis,* Vol. 17, 1980, pp. 883–893.

[5] Brown, R. M., *Emitter Location Using Bearing Measurements from a Moving Platform,* NRL Report 8483, Naval Research Laboratory, Washington, DC, June 1981.

[6] Poisel, R. A., *Introduction to Communication Electronic Warfare Systems,* Norwood, MA: Artech House, 2002, pp. 384–388.

[7] Sage, A. P., and J. L. Melsa, *Estimation Theory with Applications to Communications and Control,* New York: McGraw Hill, 1971, pp. 244–245.

[8] Poirot, J. L., and M. S. Smith, "Moving Emitter Classification," *IEEE Transactions on Aerospace and Electronic Systems,* Vol. AES-12, No. 2, March 1976, pp. 255–269.

[9] Poirot, J. L., and G. V. McWilliams, "Application of Linear Statistical Models to Radar Location Techniques," *IEEE Transactions on Aerospace and Electronic Systems,* Vol. AES-10, No. 6, November 1974, pp. 830–834.

[10] Pages-Zamora, A., J. Vidal, and D. Brooks, "Closed-Form Solution for Positioning Based on Angle of Arrival Measurements," PIMRC 2002.

[11] Rao, K. D., and D. C. Reddy, "A New Method for Finding Electromagnetic Emitter Location," *IEEE Transactions on Aerospace and Electronic Systems,* Vol. 30, No. 4, October 1994, pp. 1081–1085.

[12] Kalman, R. E., "A New Approach to Linear Filtering and Prediction Problems," *Transactions of the ASME—Journal of Basic Engineering,* 82 (Series D): 35–45.

[13] Aidala, V. J., "Kalman Filter Behavior in Bearing-Only Tracking Applications," *IEEE Transactions on Aerospace and Electronic Systems,* Vol. 15, No. 1, January 1979, pp. 29–39.

[14] Spingarn, K., "Passive Position Location Estimation Using the Extended Kalman Filter," *IEEE Transactions on Aerospace and Electronic Systems,* Vol. AES-23, No. 4, July 1987, pp. 558–567.

[15] Julier, S. J., and J. K. Uhlmann, "A New Extension of the Kalman Filter to Nonlinear Systems," *Proceedings of AeroSense: The 11th International Symposium on Aerospace/Defense Sensing, Simulation, and Controls,* 1997.

[16] Elsaesser, D. S., and R. G. Brown, "The Discrete Probability Density Method for Emitter Geo-Location (U)," Defense Research & Development Canada, Ottawa Technical Memorandum 2003-068, June 2003.

[17] Simon, M. K., and M. S. Alouini, *Digital Communication over Fading Channels,* New York: Wiley, 2000, p. 51.

[18] Paradowski, L., "Unconventional Algorithm for Emitter Location in Three-Dimensional Space Using Data from Two-Dimensional Direction Finding," *Proceedings of the IEEE 1994 National Aerospace and Electronic Systems Conference, NAECON 1994,* May 23–27, 1994, pp. 246–250.

[19] Levanon, N., "Interferometry against Differential Doppler: Performance Comparison of Two Emitter Location Airborne Systems," *IEE Proceedings,* Vol. 136, Pt. F., No. 2, April 1989, pp. 70–74.

[20] Gavish, M., and A. J. Weiss, "Performance Analysis of Bearing-Only Target Location Algorithms," *IEEE Transactions on Aerospace and Electronic Systems,* Vol. 28, No. 3, July 1992, pp. 817–828.

[21] Stansfield, R. G., "Statistical Theory of DF Fixing," *Journal of IEE,* 14, Part IIIA, 15, 1947, pp. 762–770.

[22] Poisel, R. A., *Introduction to Communication Electronic Warfare Systems,* Norwood, MA: Artech House, 2002, pp. 381–383.

[23] Gething, P. J. D., *Radio Direction Finding and Resolution of Multicomponent Wave-Fields,* London: Peter Peregrinus, Ltd, 1978, pp. 271–275.

[24] Fu, S. J., J. L. Vian, and D. L. Grose, "Determination of Ground Emitter Location," *IEEE Aerospace and Electronic Systems Magazine,* December 1988, pp. 15–18.

[25] Hammel, S. E., and V. J. Aidala, "Observability Requirements for Three-Dimensional Tracking via Angle Measurements," *IEEE Transactions on Aerospace and Electronic Systems,* Vol. AES-21, No. 2, March 1985, pp. 200–207.

[26] Poisel, R. A., *Introduction to Communication Electronic Warfare Systems,* Norwood, MA: Artech House, 2002, Chapter 11.

[27] Torrieri, D. J., "Statistical Theory of Passive Location Systems," *IEEE Transactions on Aerospace and Electronic Systems,* Vol. AES-20, No. 2, March 1984, pp. 183–198.

[28] Gavish, M., and E. Fogel, "Effect of Bias on Bearing-Only Target Location," *IEEE Transactions on Aerospace and Electronic Systems,* Vol. 26, No. 1, January 1990, pp. 22–26.

[29] McCabe, H., and M. Al-Samara, "Application of a Sensor Fusion Algorithm to Nonlinear Data Containing Nonstationary, Multiplicative Noise of Unknown Distribution," *Proceedings of the 33rd Conference on Decision and Control,* Lake Buena Vista, FL., December 1994, pp. 1205–1206.

[30] McCabe, H., "Minimum Trace Fusion of *N* Sensors with Arbitrarily Correlated Sensor-to-Sensor Fusion," IFAC International Symposium on Distributed Intelligence Systems, Arlington, VA, August 1991, pp. 229–234.

[31] Manolakis, D. E., and M. E. Cox, "Effect in Range Difference Position Estimation Due to Stations' Position Errors," *IEEE Transactions on Aerospace and Electronic Systems,* Vol. 34, No. 1, January 1998, pp. 328–334.

[32] Wang, Y., K. Ikeda, and K. Nakayama, "Computational Complexity Reduction of Predictor Based Least-squared Algorithm and Its Numerical Property," *Proceedings of the International Conference on Information, Communications, and Signal Processing, ICICS 1997,* Singapore, Sept. 9–12, 1997, pp. 1203–1207.

[33] Rao, Y. N., and J. C. Principe, "Efficient Total Least-squared Method for System Modeling Using Minor Component Analysis," *Proceedings of 2002 IEEE Workshop on Neural Networks for Signal Processing,* Sept. 4–6, 2002, pp. 259–268.

[34] Mendel, J. M., "Computational Requirements for a Discrete Kalman Filter," *IEEE Transactions on Automatic Control,* Vol. AC-16, No. 6, December 1971, pp. 748–758.

Chapter 3

Quadratic Position-Fixing Methods

Techniques for determining PF estimates of targets based on measuring the TOA, TDOA, differential frequency [otherwise known as differential Doppler (DD)], and RD are discussed in this chapter.

When the TOA or the time difference of arrival (denoted here by τ) measured at two or more widely dispersed sensors are used to geolocate emitters, then quadratic LOPs result. Generally, the intersection of these LOP curves is taken as the estimated emitter location. Calculation of this intersection can be tedious and complicated, requiring a search over a large parameter space.

The principal advantages of these forms of processing are the following:

- Frequently a single antenna per sensor is required as opposed to an array for interferometric and similar processing methods that rely on the intersection of LOBs to determine the PF.
- Normally, higher precision and more accuracy can be obtained with quadratic processing.

On the other hand, there is one primary drawback:

- Typically, preprocessed data samples are required for nonpulsed modulations. This requires interlinking data links with considerable bandwidth between sensor platforms or the sensor platform and some processing site. This is not normally a problem in pulsed signal environments where signal events can be identified (for example, the leading edge of a pulse), such as radar geolocation or digital data signals.

A technique that yields the PF in closed form using TDOA measurements was presented in [1]. Presented here are some alternative algorithms. Two of these algorithms would be expected to perform better with noisy or erroneous measurements because of their iterative search behavior.

This chapter begins with a presentation of some TDOA techniques of PF estimation. After a presentation of the definitions involved, a nonlinear least-squared technique is discussed. This is followed by a presentation of estimation of the TDOA from phase data collected at widely separated sensors. Next, a discussion of TDOA measurement accuracy is included. The independent variables in this discussion are the SNR and integration time, as these two parameters are the primary determining factors of TDOA estimation performance. Three algorithms are presented next and compared for TDOA estimation when noise is included. The effects of GDOP are present in TDOA estimation, just as they were for triangulation in Chapter 2. The last topics discussed in the TDOA section are these GDOP effects.

The second section in this chapter talks about DD methods for PF estimation. After basic definitions are presented, a brief discussion of measurement accuracy is included. This is followed by a section in which a maximum likelihood algorithm is given, along with performance estimates. After a presentation of the *cross ambiguity function* (CAF), which uses both TDOA as well as DD data, a section is included that discusses the very practical case of estimating the DD of a sinusoid in AWGN using phase data.

The last major topic presented in this chapter is PF estimation using range data. After definitions are presented, three least-squared distance error algorithms are discussed and compared—the spherical interpolation method, the spherical intersection method, and the plane intersection method. Lastly, an RD method for estimation of the PF using mathematical constructs from Grassman algebra, called feasible bivectors, is presented.

3.1 TDOA Position-Fixing Techniques

3.1.1 Introduction

The time of arrival measured at several sensors can be used for geoposition determinations. Although these arrival times can be used directly for this purpose, described in this chapter are techniques for calculating PFs using the time differences of the times of arrival at several sensors compared with a reference senor.

TOA techniques can also be, and have been for years, used for geoposition calculations. Navigation systems such as Loran use TOA to determine the location of aircraft and ships at sea. It has been shown by Rusu [2] that TDOA and TOA processing are exactly equivalent. It has been shown by Shin and Sung [3] that the error characteristics of TOA and TDOA, as manifest in the *dilution of precision*

(DOP), are exactly equivalent. We will be concerned only with TDOA processing here.

3.1.2 Time Difference of Arrival

The geometry is shown in Figure 3.1, where, for simplicity, only two receiving systems are shown. Since the transmitter and/or receiving systems can be elevated, in general r_1 and r_2 are slant ranges between the transmitter and the sensor systems. Initially here, only two dimensions will be considered, however.

In a constant velocity medium, these distances can be expressed as

$$r_i = ct_i, \qquad i = 1,2 \tag{3.1}$$

where c is the speed of propagation of the signal, normally for communication signals assumed to be the speed of light, and t_i is the time between when the signal leaves the transmitter and when it arrives at the sensor.

The τ is the time difference between when the signal arrives at one receiving site and the other:

$$\tau = t_2 - t_1 = \frac{r_2}{c} - \frac{r_1}{c} = \frac{1}{c}(r_2 - r_1) \tag{3.2}$$

From Figure 3.1,

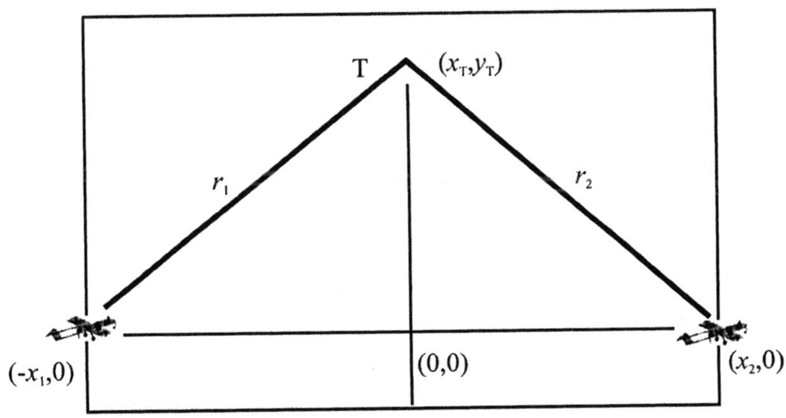

Figure 3.1 TDOA computation scenario.

$$r_i = \sqrt{(x_{\rm T} - x_i)^2 + y_{\rm T}^2}, \qquad i = 1,2 \tag{3.3}$$

so that

$$\begin{aligned}\Delta r = r_2 - r_1 &\triangleq h(x_{\rm T}, y_{\rm T}, \mathbf{x}_{\rm R}, \mathbf{y}_{\rm R}) \\ &= \sqrt{(x_{\rm T} - x_1)^2 + y_T^2} - \sqrt{(x_{\rm T} - x_2)^2 + y_{\rm T}^2}\end{aligned} \tag{3.4}$$

where $\mathbf{x}_{\rm R}$ and $\mathbf{y}_{\rm R}$ are vectors of the x- and y-coordinates of the sensors. Let d denote the length of the baseline, which is the distance between the sensors given by

$$d = x_1 + x_2 \tag{3.5}$$

This can be rewritten in the form of a hyperbola as

$$\frac{x^2}{a} - \frac{y^2}{b} = \frac{x^2}{(\Delta r)^2 / d^2} - \frac{y^2}{d^2/4 - (\Delta r)^2/4} = 1 \tag{3.6}$$

The resulting LOPs are called *isochrones,* since they characterize points in space where τ is a constant. The target can lie at any point on the isochrone.

Denote (X, Y) as the global coordinates for sensor positions that are located off the axis as in Figure 3.1. If (x, y) are the local coordinates, then

$$\begin{bmatrix} X \\ Y \end{bmatrix} = \begin{bmatrix} X_0 \\ Y_0 \end{bmatrix} + \begin{bmatrix} \cos\alpha & -\sin\alpha \\ \sin\alpha & \cos\alpha \end{bmatrix} \begin{bmatrix} x \\ y \end{bmatrix} \tag{3.7}$$

where

$$X_0 = \frac{X_i + X_j}{2} \qquad Y_0 = \frac{Y_i + Y_j}{2} \tag{3.8}$$

$$D = \sqrt{(Y_i - Y_j)^2 + (X_i - X_j)^2} \tag{3.9}$$

$$\begin{bmatrix} x \\ y \end{bmatrix} = \begin{bmatrix} \cos\alpha & \sin\alpha \\ -\sin\alpha & \cos\alpha \end{bmatrix} \begin{bmatrix} X - X_0 \\ Y - Y_0 \end{bmatrix} \tag{3.10}$$

and

$$\alpha = \tan^{-1}\left(\frac{Y_i - Y_j}{X_i - X_j}\right) \tag{3.11}$$

The τ contour for the scenario shown in Figure 3.1 is shown in Figure 3.2 when d = 80 km. The τ varies between about –130 μsec to +130 μsec. Also shown is a constant τ curve; in this case, τ is 90 μsec. The intersection of these curves forms a hyperbola as shown. As visualized from above, several of these curves are shown in Figure 3.3.

Let N denote the number of sensors. Then there are

$$K = \binom{N}{2} \tag{3.12}$$

sensor pairs, or hyperbolas, available for geolocation.

3.1.3 Calculating the PF with TDOAs

Consider the three-dimensional geometry shown in Figure 3.4. It will be assumed that, without loss of generality, a reference sensor is placed at the origin. It is also assumed without loss of generality that the reference sensor is sensor 1. This section follows [4] closely; a similar derivation was documented by Mellen et al. [5].

Measurement at the ith sensor is given by the expression

$$(x_\mathrm{T} - x_i)^2 + (y_\mathrm{T} - y_i)^2 + (z_\mathrm{T} - z_i)^2 = c^2(\tau_i + t_0) \tag{3.13}$$

where t_0 is the unknown arrival time at the reference sensor placed at the origin. The target is located at $\mathbf{x} = \begin{bmatrix} x_\mathrm{T} & y_\mathrm{T} & z_\mathrm{T} \end{bmatrix}^\mathrm{T}$. At the reference site,

$$x_\mathrm{T}^2 + y_\mathrm{T}^2 + z_\mathrm{T}^2 = (ct_0)^2 = r_0^2 = \|\mathbf{x}\|_2^2 = \mathbf{x}^\mathrm{T}\mathbf{x} \tag{3.14}$$

where $\|\mathbf{x}\|_2$ denotes the L_2 norm of $\mathbf{x}$. Combining (3.13) and (3.14) yields

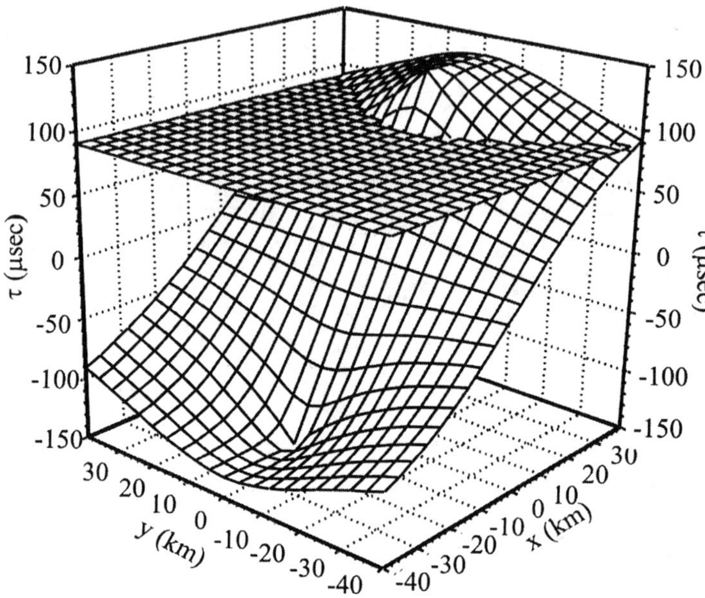

Figure 3.2 The intersection of the τ curve with the plane defined by a constant τ curve of 90 μsec forms a hyperbolic curve. (*Source:* [1], © 2002 Horizon House. Reprinted with permission.)

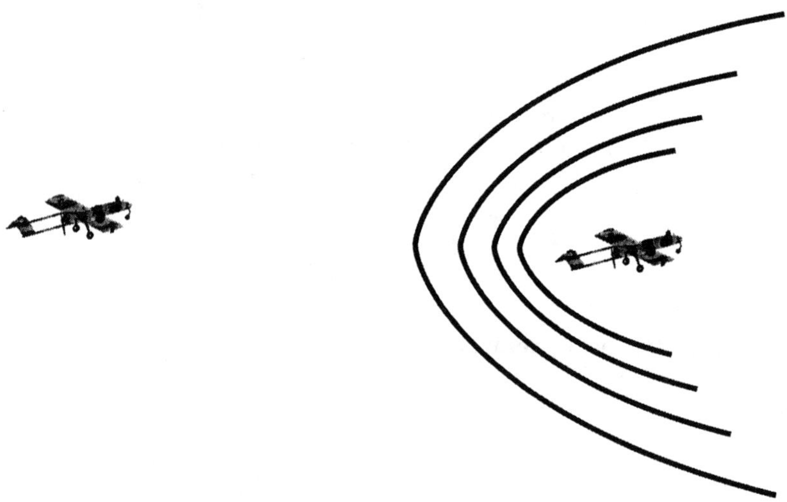

Figure 3.3 TDOA isochrones with no error. (*Source:* [1], © 2002 Horizon House. Reprinted with permission.)

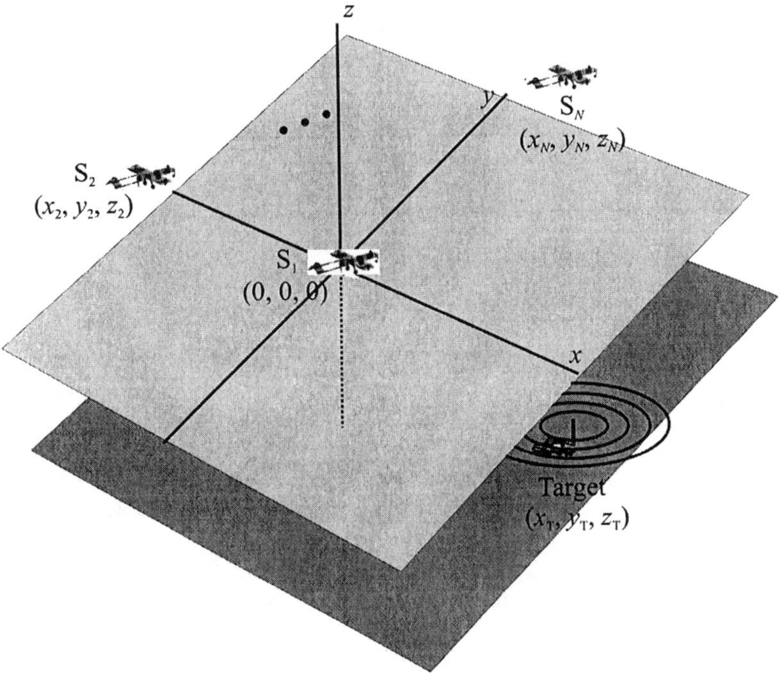

Figure 3.4 TDOA geometry.

$$\begin{bmatrix} x_i & y_i & z_i \end{bmatrix} \begin{bmatrix} x_T \\ y_T \\ z_T \end{bmatrix} + c\tau_i \sqrt{x_T^2 + y_T^2 + z_T^2}$$

$$= -\frac{1}{2}c^2\tau_i^2 + \frac{1}{2}(x_i^2 + y_i^2 + z_i^2), \qquad i = 1, 2, \ldots, M \tag{3.15}$$

In matrix form, (3.15) is represented as

$$\mathbf{Ax} + \mathbf{c}\|\mathbf{x}\|_2 = \mathbf{M} \tag{3.16}$$

where

$$\mathbf{A} = \begin{bmatrix} \mathbf{x}_1^{\mathrm{T}} \\ \mathbf{x}_2^{\mathrm{T}} \\ \vdots \\ \mathbf{x}_M^{\mathrm{T}} \end{bmatrix} \in \Re^{N\times 3}$$

$$\mathbf{x}_i = [x_i \quad y_i \quad z_i]^{\mathrm{T}}$$

$$\mathbf{c} = c\boldsymbol{\tau} \in \Re^{N\times 1}$$

$$\boldsymbol{\tau} = [\tau_1 \quad \tau_2 \quad \cdots \quad \tau_N]^{\mathrm{T}}$$

$$\mathbf{M} = \left[\frac{1}{2}(\mathbf{A}\mathbf{A}^{\mathrm{T}} - c^2\boldsymbol{\tau}\boldsymbol{\tau}^{\mathrm{T}})\right]_{ii}, \qquad i = 1,2,\ldots,N \in \Re^{N-1\times 1}$$

$\Re^{M\times N}$ denotes the set of M-tuples, each consisting of N real numbers.

In general (3.16) will be overdetermined in that there are more measurements than the minimum number required; that is $\mathbf{A} \in \Re^{M\times N}$, where $M > N$. Equation (3.16) can be rearranged as

$$\mathbf{A}\mathbf{x} = \mathbf{M} - \mathbf{c}\|\mathbf{x}\|_2 = \mathbf{f}(p) \tag{3.17}$$

The vector $\mathbf{f} \in \Re^{M\times 1}$ has the parameter p such that, as discussed in Section 2.3, for any p, the pseudoinverse of $\mathbf{A}$ given by

$$\mathbf{A}^* = (\mathbf{A}^{\mathrm{T}}\mathbf{A})^{-1}\mathbf{A}^{\mathrm{T}} \tag{3.18}$$

solves (3.17) in the least-squared sense; thus

$$\mathbf{x}^* = \mathbf{A}^*(\mathbf{M} - \mathbf{c}p) \tag{3.19}$$

solves (3.17) such that the residual

$$\varepsilon = \left\|\mathbf{A}\mathbf{x}^* - (\mathbf{M} - \mathbf{c}p)\right\|_2 \tag{3.20}$$

is minimized.

The measurements are used in the computation of the scalar equation

$$(\mathbf{c}^{\mathrm{T}}\boldsymbol{\Phi}\mathbf{c} - 1)p^2 - 2\mathbf{M}^{\mathrm{T}}\boldsymbol{\Phi}\mathbf{c}p + \mathbf{M}^{\mathrm{T}}\boldsymbol{\Phi}\mathbf{M} = 0 \tag{3.21}$$

where

$$\mathbf{\Phi} = \mathbf{A}(\mathbf{A}^{\mathrm{T}}\mathbf{A}\mathbf{A}^{\mathrm{T}}\mathbf{A})^{-1}\mathbf{A}^{\mathrm{T}} \tag{3.22}$$

With noise present, the measurements are corrupted so that

$$\mathbf{Ax} = \mathbf{f}(p) + \mathbf{n} \tag{3.23}$$

with

$$\mathcal{E}\{\mathbf{n}\} = 0 \qquad \mathcal{E}\{\mathbf{n}\mathbf{n}^{\mathrm{T}}\} = \mathbf{Q} \tag{3.24}$$

The pseudoinverse in this case is

$$\mathbf{A}^{*} = (\mathbf{A}^{\mathrm{T}}\mathbf{Q}^{-1}\mathbf{A})^{-1}\mathbf{A}^{\mathrm{T}}\mathbf{Q} \tag{3.25}$$

If the noise is uncorrelated among sensors with equal variances given by σ_n^2, then

$$\mathbf{Q} = \sigma_n^2\mathbf{I} \tag{3.26}$$

In this case, (3.22) must be changed to account for the noise as

$$\mathbf{\Phi} = (\mathbf{A}^{*})^{-1}\mathbf{A}^{*}[(\mathbf{A}^{*})^{\mathrm{T}}\mathbf{Q}^{-1}\mathbf{A}^{*}(\mathbf{A}^{*})^{\mathrm{T}}\mathbf{Q}^{-1}\mathbf{A}^{*}]^{-1}(\mathbf{A}^{*})^{\mathrm{T}}\mathbf{Q}^{-1} \tag{3.27}$$

The solution of (3.21) determines p, which is equal to r_0, the range from the reference sensor to the target. Thus,

$$p = r_0 = \|\mathbf{x}\|_2 = \sqrt{\mathbf{x}^{\mathrm{T}}\mathbf{x}} \tag{3.28}$$

and is used to compute the estimate of the geolocation of the target as

$$\mathbf{x}^{*}(r_0) = \begin{bmatrix} x_{\mathrm{T}} & y_{\mathrm{T}} & z_{\mathrm{T}} \end{bmatrix}^{\mathrm{T}} = \mathbf{A}^{*}(\mathbf{M} - \mathbf{c}r_0) \tag{3.29}$$

The Taylor series gradient descent algorithm described in Chapter 1 is one of the most accurate methods for estimating a PF. If the PF computed as indicated here is not accurate enough, the PF can at least be used as a starting point for the descent algorithm to improve upon the accuracy. As mentioned in Chapter 1, the gradient descent algorithm can use any type of measurement.

3.1.4 Nonlinear Least-Squared

The problem to solve is to find the optimum solution to (3.4) in generalized coordinates, which is a nonlinear minimization problem. Note that this expression could be overdetermined by the measurements. In two-dimensions, the nonlinear least-squared estimate of the target location (x_T, y_T) is given by

$$(\hat{x}_\text{T}, \hat{y}_\text{T}) = \underset{(x,y)}{\text{argmin}} \sum_{i>j} [\Delta r_{i,j} - h(x_\text{T}, y_\text{T}, x_i, y_i, x_j, y_j)]^2 \quad 1 \le i < j \le N \quad (3.30)$$

Denoting the target location being sought by $\mathbf{P} = \begin{bmatrix} x_\text{T} & y_\text{T} \end{bmatrix}^\text{T}$, then this minimization becomes

$$\hat{\mathbf{P}} = \underset{\mathbf{P}}{\text{argmin}} [\boldsymbol{\Delta}\mathbf{r} - \mathbf{h}(\mathbf{P})]^\text{T} \mathbf{R}^{-1} [\boldsymbol{\Delta}\mathbf{r} - \mathbf{h}(\mathbf{P})] \quad (3.31)$$

where

$$\boldsymbol{\Delta}\mathbf{r} = [\Delta r_{1,2} \quad \Delta r_{2,3} \quad \cdots \quad \Delta r_{N-1,N}] \quad (3.32)$$

and

$$\mathbf{R} = \mathbf{Cov}(\boldsymbol{\Delta}\mathbf{r}) \quad (3.33)$$

Assuming the calculated position is close to the true position, $\mathbf{P}_0$, then $\boldsymbol{\Delta}\mathbf{r} = \mathbf{h}(\mathbf{P}_0) + \boldsymbol{\varepsilon}$, and if $\mathbf{R} = \mathbf{Cov}(\boldsymbol{\varepsilon})$, then, based on a Taylor series expansion and retaining only the linear terms,

$$\mathbf{Cov}(\hat{\mathbf{P}}) = [\dot{\mathbf{h}}_\text{p}(\mathbf{P}_0)]^\text{H} \mathbf{R} \{[\dot{\mathbf{h}}_\text{p}(\mathbf{P}_0)]^\text{H}\}^\text{T} \quad (3.34)$$

where H denotes Hermitian, or conjugate transpose, and $\dot{\mathbf{h}}$ denotes differentiation of $\mathbf{h}$ with respect to $\mathbf{x}$. When $\boldsymbol{\varepsilon} \sim \mathcal{N}(\mathbf{0}, \sigma^2)$, then (3.34) describes the CRLB.

3.1.5 Estimating the TDOA from Phase Data

Given the stochastic signal transmitted by a target $s(t)$, the signals received at two geographically dispersed receive sensors $s_j(t)$ and $s_k(t)$ are given by

$$s_j(t) = s(t) + n_j(t) \quad (3.35)$$

$$s_k(t) = As(t - \tau_{jk}) + n_j(t) \tag{3.36}$$

where A is an attenuation factor and τ_{jk} is the relative time delay between the signals received at the two sensors. The cross-correlation function of these two signals is given by

$$\begin{aligned}
r_{jk}(\tau) &= \mathcal{E}\{s_j(t)s_k(t-\tau)\} \\
&= \mathcal{E}\{[s(t)+n_j(t)][As(t-\tau-\tau_{jk})+n_k(t)\} \\
&= \mathcal{E}\{[As(t)s(t-\tau-\tau_{jk})+An_j(t)s(t-\tau-\tau_{jk}) \\
&\qquad +s(t)n_k(t)+n_j(t)n_k(t)\} \\
&= A\mathcal{E}\{[s(t)s(t-\tau-\tau_{jk})\}+A\mathcal{E}\{n_j(t)s(t-\tau-\tau_{jk})\} \\
&\qquad +\mathcal{E}\{s(t)n_k(t)\}+\mathcal{E}\{n_j(t)n_k(t)\} \\
&= A\mathcal{E}\{s(t)s(t-\tau-\tau_{jk})\} \\
&= Ar_{ss}(\tau-\tau_{jk})
\end{aligned} \tag{3.37}$$

where $r_{ss}(\tau)$ is the autocorrelation function of s(t). Two characteristics of correlation functions are $|r(\tau)| \le 1$ and $\max[r(\tau)] = r(0) = 1$. Therefore, $r_{jk}(\tau)$ peaks at $\tau = \tau_{jk}$, as shown in Figure 3.5.

The Fourier transform of (3.37) yields the *power spectral density* (psd) function as

$$P_{jk}(f) = \int_{-\infty}^{\infty} r_{jk}(\tau)e^{-j2\pi f\tau}d\tau$$

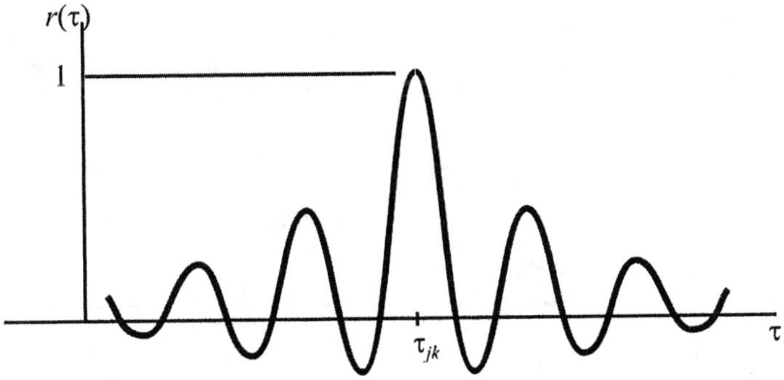

Figure 3.5 Typical autocorrelation function.

$$P_{jk}(f)=\int_{-\infty}^{\infty}Ar_{ss}(\tau-\tau_{jk})e^{-j2\pi f\tau}d\tau$$

Making the substitution $\zeta=\tau-\tau_{jk}$ yields

$$\begin{aligned}P_{jk}(f)&=\int_{-\infty}^{\infty}Ar_{ss}(\zeta)e^{-j2\pi f(\zeta+\tau_{jk})}d\tau\\&=\int_{-\infty}^{\infty}Ar_{ss}(\zeta)e^{-j2\pi f\zeta}e^{-2\pi f\tau_{jk}}d\tau\\&=Ae^{-\tau_{jk}}\int_{-\infty}^{\infty}r_{ss}(\zeta)e^{-j2\pi f\zeta}d\tau\\&=AP_{ss}(f)e^{-j2\pi f\tau_{jk}}\end{aligned}\tag{3.38}$$

where $P_{ss}(f)$ is the psd of $s(t)$.

Thus, the time delay generates a phase difference in the frequency domain given by

$$\phi_{jk}(f)=2\pi f\tau_{jk}\tag{3.39}$$

and the time delay (the TDOA) can be estimated by estimating this phase function. In the absence of noise, this phase function is linearly dependent on the frequency, as illustrated in Figure 3.6. The estimated time delay is given by the slope of this linear function divided by 2π.

With noise present, the phase curve is not linear, but a linear regression estimate can be obtained from the measured values of phase. It is assumed that the phase is measured at discrete times, indexed by i. The linear regression was derived in Section 2.2 and the optimum estimates of the linear coefficients in the

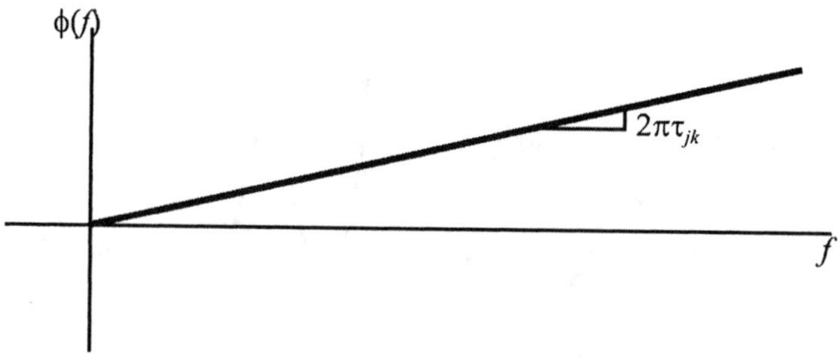

Figure 3.6 Linear phase function caused by the TDOA at two sensors.

MLSE sense are given by (2.31). Those results yield the value of the TDOA as [6]

$$\hat{\tau} = \frac{1}{2\pi} \frac{\sum_{i=1}^{N/2} f_i \hat{\phi}_{jk}(f_i)}{\sum_{i=1}^{N/2} f_i^2} \tag{3.40}$$

with the variance given by

$$\hat{\sigma}^2_{\hat{\tau}_{jk}} = \left[8\pi^2 N_s \sum_{i=1}^{N/2} \frac{f_i^2 \hat{C}_{jk}(f_i)}{1 - \hat{C}_{jk}(f_i)} \right]^{-1/2} \tag{3.41}$$

where it has been assumed that the total measurement time has been divided into N_s segments for the analysis. $\hat{C}_{jk}(f_i)$ is the estimate of the coherence function of the two sensors at f_i given by

$$\hat{C}_{jk}(f_i) = \frac{\left| \hat{P}_{jk}(f_i) \right|^2}{\hat{P}_{jj}(f_i) \hat{P}_{kk}(f_i)} \tag{3.42}$$

3.1.6 TDOA Measurement Accuracy

When errors are taken into consideration, the TDOA isochrones are no longer clean functions, as illustrated in Figure 3.3, but form regions or areas within which the target should lie, as illustrated in Figure 3.7.

All of the methods of computing the geolocation of targets with measurement of the TDOAs are subject to errors in the measurements. The accuracy of TDOA measurements is discussed in this section in terms of the standard deviation. The noise and measurement errors are the two primary sources of error. Included here are the effects of noise that determine the SNR at the receiver. Measurement errors are systematic and unique to each implementation. They will not be discussed here.

3.1.6.1 Expected Accuracy of TDOA PF estimates

As indicated in [1], the Cramer-Rao bound on parameter estimation is a frequently used measure of how well such a parameter can be measured, as long as the

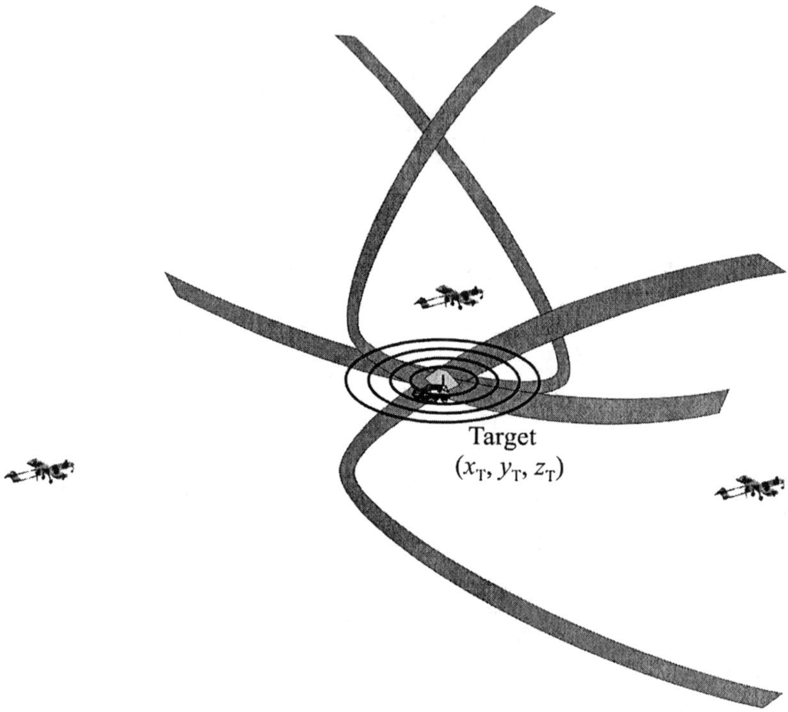

Figure 3.7 TDOA isochrones with errors included. The target indicated location could be anywhere in the light area.

parameter is unbiased. Under some reasonable assumptions it represents the best that can be obtained under those assumptions. The Cramer-Rao bound for estimating the time of arrival of a signal at a sensor is given by [7]

$$\sigma_\tau = \frac{1}{\beta}\frac{1}{\sqrt{BT\gamma}} \tag{3.43}$$

where B is the noise bandwidth of the receivers, T is the integration time, and γ is the effective input SNR at the two sensor sites.

The RMS radian frequency is given by β, which is a measure of the bandwidth of the signal and is given by

$$\beta = 2\pi \left[\frac{\int_{-\infty}^{\infty} f^2 |S(f)|^2 \, df}{\int_{-\infty}^{\infty} |S(f)|^2 \, df} \right]^{1/2} \tag{3.44}$$

where $S(f)$ is the spectrum of the signal. For an ideal signal with sharp edges in the frequency spectrum, $\beta = 2\pi B/\sqrt{3}$ (where B is in hertz) for example.

Variable γ is a composite SNR at the two sensors. If γ_i and γ_j are the SNRs at the two sensors, then γ is given by

$$\frac{1}{\gamma} = \frac{1}{2}\left[\frac{1}{\gamma_i} + \frac{1}{\gamma_j} + \frac{1}{\gamma_i \gamma_j}\right] \tag{3.45}$$

3.1.6.2 Low SNR

Quazi compared the theoretical measurement accuracy of several TDOA computation models [8]. For low SNR levels, the standard deviation computed with the Knapp and Carter [9], Schultheiss [10], Hahn [11], and Tomilinson and Sorokowsky [12] techniques all evaluated to

$$\sigma_\tau \geq \sqrt{\frac{3}{8\pi^2 T}} \frac{1}{\delta} \frac{1}{\sqrt{f_2^3 - f_1^3}} \tag{3.46}$$

where T is the integration time, δ is the SNR, and it was assumed that both the signal and noise were constant levels over the bandwidth specified by f_1 to f_2. This function is illustrated in Figures 3.8 and 3.9 for a 25-kHz bandwidth for several values of T. This expression can be represented in terms of the center frequency f_0 and bandwidth $W = f_2 - f_1$ as

$$\sigma_\tau \approx \sqrt{\frac{1}{8\pi^2}} \frac{1}{\delta} \frac{1}{\sqrt{TW}} \frac{1}{f_0} \frac{1}{\sqrt{1 + \frac{W^2}{12 f_0^2}}} \tag{3.47}$$

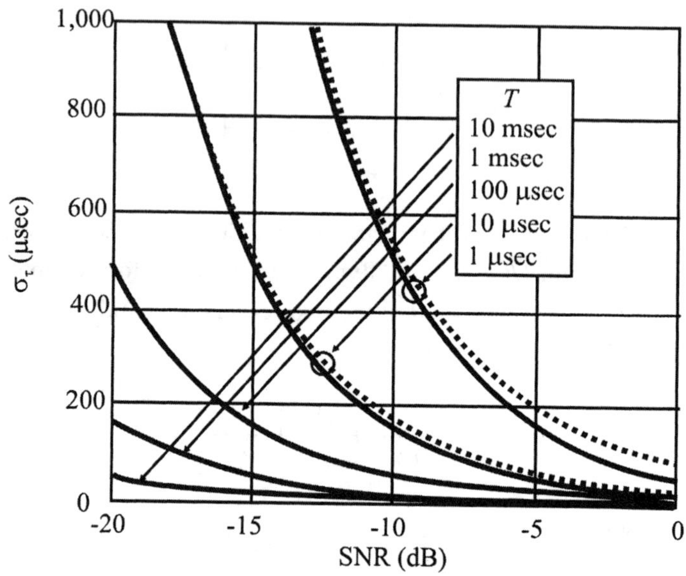

Figure 3.8 TDOA standard deviation for low SNRs and short integration times. The dotted lines are (3.43) while the solid lines are (3.46). For $T > 10\ \mu$sec, the differences in the lines are not discernable.

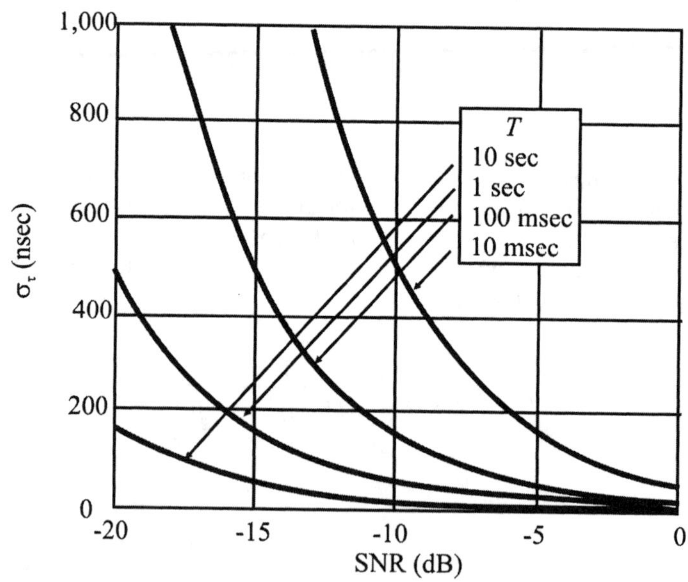

Figure 3.9 TDOA standard deviations for longer integration times, $W = 25$ kHz. Both (3.43) and (3.46) are plotted with (3.43) given by dotted lines and (3.46) by solid lines, but, as with Figure 3.8, the differences in the curves is not discernable because $T > 10\ \mu$sec.

$$\sigma_\tau \approx \sqrt{\frac{1}{8\pi^2}\frac{1}{\delta}\frac{1}{\sqrt{TW}}\frac{1}{f_0}\frac{1}{\sqrt{1+\frac{W^2}{12 f_0^2}}}} \tag{3.48}$$

3.1.6.3 High SNR

The standard deviation when the SNR is high, SNR >> 1, is given by

$$\sigma_\tau \approx \sqrt{\frac{3}{4\pi^2 T}\frac{1}{\sqrt{\delta}}\frac{1}{\sqrt{f_2^3 - f_1^3}}} \tag{3.49}$$

which is the same as (3.46) with the square root of the SNR in the denominator rather than just the SNR. This function is plotted in Figures 3.10 and 3.11.

3.1.7 TDOA PFs with Noisy Measurements

Gustafsson and Gunnarsson [13] presented three algorithms for determining the PF of a target based on TDOA measurements that may contain measurement errors. Those algorithms are discussed and compared in this section.

3.1.7.1 Isochrone Intersection Algorithm

The first of these computes the intersection point of each pair of isochrones. With K pairs of sensors, this results in

$$\binom{K}{2} = \frac{K!}{2!(K-2)!} \tag{3.50}$$

points to estimate. The PF estimate is given by the point where all $\binom{K}{2}$ of these isochrones intersect (or perhaps point where the largest number of them intersect). Each pair of isochrones can have zero, one, or two intersections. Determining such intersections can be difficult. The authors claim that this approach leads to inferior results compared to the other two algorithms when the measurements are corrupted with noise. With noise present, it could be that no more than two isochrones intersect at any point.

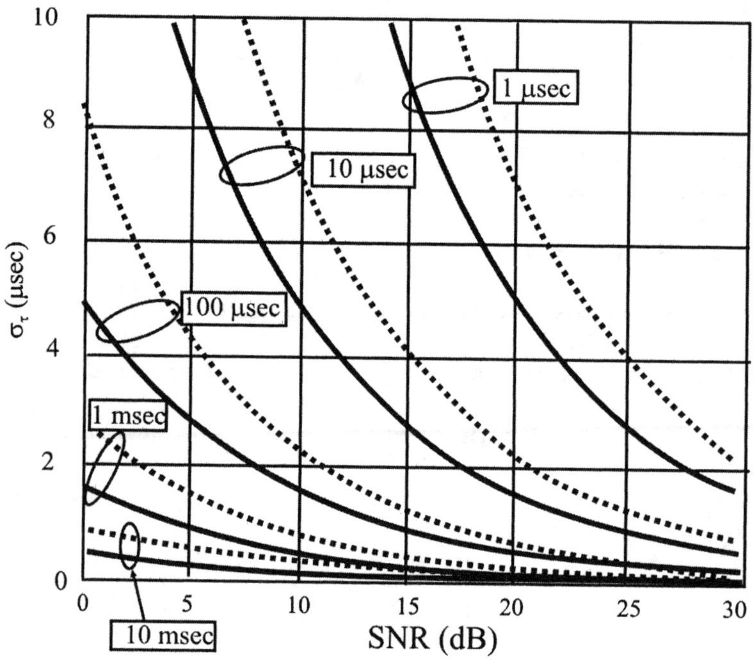

Figure 3.10 TDOA standard deviation for higher SNRs and short integration times, $W = 25$ kHz. The dotted lines are (3.43) and the solid lines are (3.49).

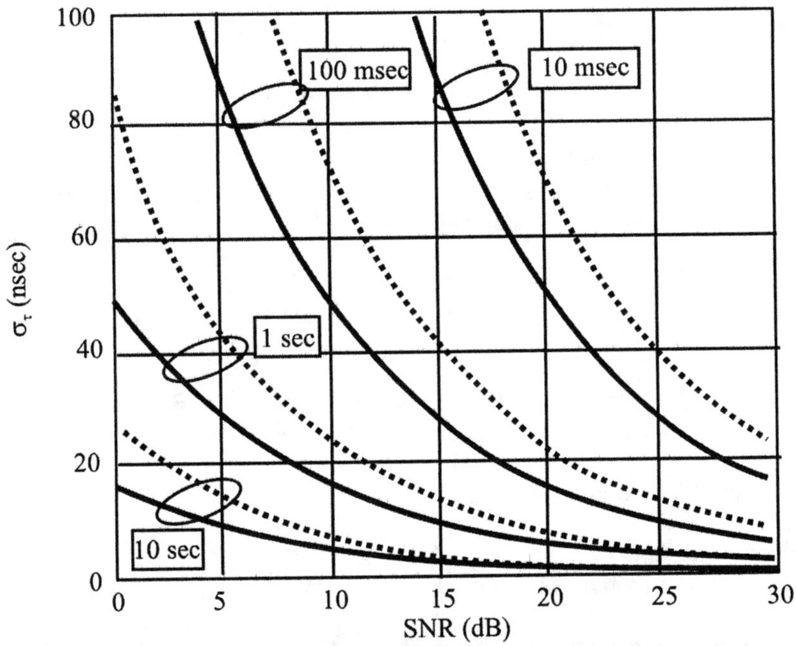

Figure 3.11 TDOA standard deviation for higher SNRs and long integration times, $W = 25$ kHz. The dotted lines are (3.43) and the solid lines are (3.49).

3.1.7.2 Stochastic Gradient Algorithm

The second algorithm applies the stochastic gradient algorithm to the nonlinear least-squared problem described above. Let $\mathbf{P} = [x_\text{T}\ y_\text{T}]^\text{T}$ denote the target position and $\hat{\mathbf{P}}^{(i)}$ the estimate of **P** on the *i*th iteration. In this algorithm,

$$\hat{\mathbf{P}}^{(i+1)} = \hat{\mathbf{P}}^{(i)} - \mu^{(i)} h'_P(\hat{\mathbf{P}}^{(i)})[\Delta\mathbf{r} - h(\hat{\mathbf{P}}^{(i)})] \tag{3.51}$$

The step size $\mu^{(i)}$ can be found using

$$\mu^{(i)} = \frac{\mu}{[\mathbf{h}'_P(\hat{\mathbf{P}}^{(i)})]^T \mathbf{h}'_P(\hat{\mathbf{P}}^{(i)})} \tag{3.52}$$

where μ is the *least-mean squared* (LMS) step size.

Step size μ is selected based on stability considerations. The sequence (3.32) will converge if and only if μ satisfies

$$0 < \mu < \frac{1}{\lambda_\text{max}} \tag{3.53}$$

where λ_max is the largest eigenvalue of the covariance matrix **R** given by (3.33). Of course, the larger the value of μ, the faster the convergence, typically.

3.1.7.3 Static Particle Filter Algorithm

The third algorithm in [13] is based on the *static particle filter* [14–16]. The particle filter is a method of implementing a recursive Bayesian filter by Monte-Carlo simulations. The a posteriori density function is represented by a set of random samples with associated weights and estimates based on these samples and weights are computed. The static particle filter algorithm is an approximation of the optimal Bayesian filter described here. Note that the Kalman filter (not the EKF) described in Section 2.9.1 is an implementation of the optimal Bayesian filter.

Given the system state equation that describes the evolution of the state sequence of the target

$$\mathbf{x}_k = \mathbf{f}_k(\mathbf{x}_{k-1}, \mathbf{n}_{k-1}) \tag{3.54}$$

where $\{\mathbf{x}_k,\ k \in \mathbb{N}\}$. The set $\{\mathbf{x}_k,\ k \in \mathbb{N}\}$ describes the state of the system at time step k, $\mathbf{f}_k : \Re^{N_x} \times \Re^{N_n} \to \Re^{N_x}$ describes the state update from $k - 1$ to k and may be nonlinear, $\{\mathbf{n}_k, k \in \mathbb{N}\}$ is an i.i.d noise process sequence, N_x and N_n are the dimensions of the state and noise vectors, respectively, and $\mathbb{N}$ is the set of natural numbers. The measurements at time k are given by

$$\mathbf{z}_k = \mathbf{h}_k(\mathbf{x}_k, \boldsymbol{\eta}_k) \tag{3.55}$$

where $\mathbf{h}_k : \Re^{N_x} \times \Re^{N_\eta} \to \Re^{N_x}$ is a (possibly nonlinear) function, $\{\boldsymbol{\eta}_k, k \in \mathbb{N}\}$ is an i.i.d. measurement noise sequence, and N_η is the dimension of that noise process. The goal is to estimate the states $\mathbf{x}_k$ based on the set of measurements up to time k: $\mathbf{z}_{1:k} = \{\mathbf{z}_i,\ i = 1, \ldots, k\}$.

The pdf $p(\mathbf{x}_k | \mathbf{z}_{1:k})$ is the desired function to be maximized to provide the best estimate of $\mathbf{x}_k$. This pdf is obtained recursively by prediction and updating. The initial pdf $p(\mathbf{x}_0 | \mathbf{z}_0)$ is assumed to be known. If the pdf $p(\mathbf{x}_{k-1} | \mathbf{z}_{1:k-1})$ is known, then (3.54) and the Chapman-Kilmogorov equation

$$p(\mathbf{x}_k | \mathbf{z}_{1:k-1}) = \int p(\mathbf{x}_k | \mathbf{x}_{1:k-1})\, p(\mathbf{x}_{k-1} | \mathbf{z}_{1:k-1})\, d\mathbf{x}_{k-1} \tag{3.56}$$

which uses the fact that (3.54) is a Markov process of order 1 so that $p(\mathbf{x}_k | \mathbf{x}_{1:k-1}, \mathbf{z}_{1:k-1}) = p(\mathbf{x}_k | \mathbf{x}_{1:k-1})$, are used to obtain the prior pdf of the state at time k. The pdf $p(\mathbf{x}_k | \mathbf{x}_{1:k-1})$ is determined from (3.54) and the noise characteristics that are assumed to be known.

The pdf is updated for time k using Bayes' rule

$$p(\mathbf{x}_k | \mathbf{z}_{1:k}) = \frac{p(\mathbf{z}_k | \mathbf{x}_k)\, p(\mathbf{x}_k | \mathbf{z}_{1:k-1})}{p(\mathbf{z}_k | \mathbf{z}_{1:k-1})} \tag{3.57}$$

where

$$p(\mathbf{z}_k | \mathbf{z}_{1:k-1}) = \int p(\mathbf{z}_k | \mathbf{x}_{1:k}) p(\mathbf{x}_k | \mathbf{z}_{1:k-1})\, d\mathbf{x}_k \tag{3.58}$$

which is a function of (3.55) and noise process $\boldsymbol{\eta}_k$.

There are four principal concepts of the particle filter algorithm:

(1) Bayesian inference;
(2) Monte Carlo samples;
(3) Importance sampling;
(4) Resampling.

Based on the above development, the Bayesian inference consists of $\mathbf{x}_k$, which is an unknown vector of random variables. The set of observations is given by $\mathbf{z}_k$, which is also a set of random variables. We wish to estimate $\mathbf{x}_k$ by knowing $\mathbf{z}_k$. The a priori pdf $p(\mathbf{x}_k)$ contains our prior knowledge of $\mathbf{x}_k$. The conditional pdf $p(\mathbf{z}_k|\mathbf{x}_{1:k})$ describes the relationship between $\mathbf{z}_k$ and $\mathbf{x}_k$. The a posteriori pdf $p(\mathbf{x}_k|\mathbf{z}_{1:k})$ represents the information about $\mathbf{x}_k$ given measurement $\mathbf{z}_k$.

The a posteriori pdf $p(\mathbf{x}_k|\mathbf{z}_{1:k})$ may be difficult or impossible to compute in closed form. Thus, Monte Carlo samples, also called *particles*, are used to approximate this pdf. Each particle has a value and an associated weight. Ideally, each particle would represent a sample from the pdf $p(\mathbf{x}_k|\mathbf{z}_{1:k})$. Since in general this pdf is not known in closed form, *importance sampling* is used. The particles are weighted with weights drawn from an importance distribution denoted $q(\mathbf{x}_{0:k}^i|\mathbf{z}_{1:k})$ for weight i. Selection of this distribution is typically based on some optimality criterion [15].

In estimation with the particle filter, most weights tend to zero except a few, which become large. The few that grow large are those that closely match the observations. Resampling is done to concentrate particles in regions where $p(\mathbf{x}_k|\mathbf{z}_{1:k})$ is larger.

These steps are shown graphically in Figure 3.12. The steps to the particle filter algorithm for computing the PF estimate based on TDOA measurements are as follows:

(1) Randomize L possible target locations as $\mathbf{P}^i$ from the a priori distribution $p(\mathbf{P}_0)$.
(2) Choose jittering constants C_R and C_Q and let the position random walk covariance $\bar{Q} = C_Q / k^2$ and jittering measurement noise $\bar{R} = R + C_\text{R} / k^2$.
(3) Iterate for $k = 1, 2, \ldots$ until $\hat{\mathbf{P}}_k$ has converged.
    a. Compute the location weights $\mathbf{w}^i$ using the likelihood

$$\mathbf{w}^i = \exp\{[\boldsymbol{\Delta}\mathbf{r} - \mathbf{h}(\mathbf{P}^i)]^\text{T} \mathbf{R}^{-1} [\boldsymbol{\Delta}\mathbf{r} - \mathbf{h}(\mathbf{P}^i)]\} \tag{3.59}$$

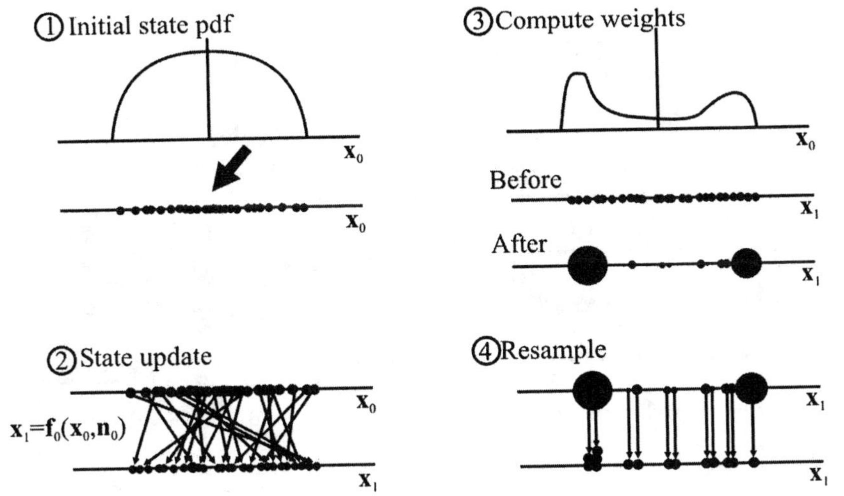

Figure 3.12 Steps in the particle filter algorithm.

where $\Delta\mathbf{r}$ is given by (3.32), $\mathbf{R}$ is given by (3.33), and $\mathbf{w}_i$ is normalized to

$$\mathbf{w}^i = \frac{\mathbf{w}^i}{\sum \mathbf{w}^i} \tag{3.60}$$

b. Compute the estimate

$$\hat{\mathbf{P}}_k = \sum_i \mathbf{w}^i \mathbf{P}^i \tag{3.61}$$

c. Resample with replacement the locations, where the probability to pick one location is proportional to its weight. After the resampling, the weights are reset $w^i = 1/L$.

d. Spread out the locations as $\mathbf{P}^i = \mathbf{P}^i + \mathbf{w}$, where $\mathbf{w} \in \mathcal{N}(0, \bar{\mathbf{Q}})$.

The idea with jittering noise in step (2) is to explore a smaller and smaller neighborhood more accurately.

A comparison example of these algorithms is shown in Figure 3.13. The receiver and target layout are shown in Figure 3.13(a), along with the results of the

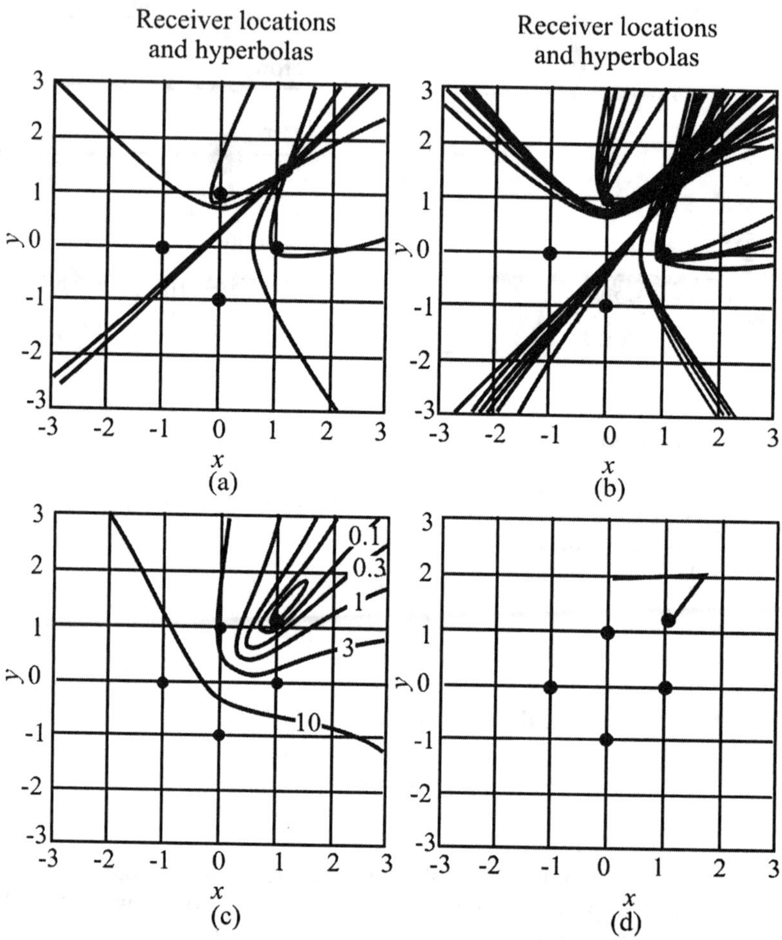

Figure 3.13 Algorithm comparison example. The test scenario is illustrated in (a), which also shows the PF estimate for the first algorithm, with noise-free hyperbolas. Panel (b) also shows the results of the stochastic gradient algorithm but with noisy data. Panel (c) is the contour plot of the nonlinear least-squared criterion $\Sigma_{i<j}[\Delta r_{ij} - h(x_T, y_T)]^2$. Panel (d) is the gradient search with the static particle filter algorithm. (*Source:* [12], © 2002 IEEE. Reprinted with permission.)

stochastic gradient algorithm with no noise. The receivers are placed at (–1, 0), (0, –1), (1, 0), and (0, 1), while the target is located at (1.2, 1.2). In the noise-free case, all six hyperbola intersect at the target location. With noise added, the results are as shown in Figure 3.13(b). There is no unique target location computed in this case. The results for the static particle filter algorithm are shown in Figure 3.13(c) with the contour plots of the nonlinear least-squared criterion:

$$\sum_{i<j}[\Delta r_{i,j} - h(x_{\mathrm{T}}, y_{\mathrm{T}})]^2 \tag{3.62}$$

The algorithm is seen to converge to the correct target location. The results for the gradient search algorithm using a normalized least mean square method with the static particle filer algorithm are illustrated in Figure 3.13(d). Again, convergence to the correct target location occurs. In this case there are no local minima and convergence to the correct target location is assured.

3.1.8 TDOA Dilution of Precision

TDOA geoposition estimating suffers from another type of error caused by long ranges from the sensor baseline. Consider Figure 3.14, where three sensors are included and the target is a considerable distance from the baselines between the sensors. The hyperbolic LOPs are nearly parallel in this region, and noise or small measurement errors can cause considerable errors in computing the geolocation. This is called GDOP or sometimes just DOP. The farther away from the sensor baselines, the worse GDOP gets.

Combining (3.13) and (3.14) yields (3.15). Expressing (3.15) in vector form,

$$\mathbf{a}_i\mathbf{x} + c\tau_i \|\mathbf{x}\|_2 = -\frac{1}{2}c^2\tau_i^2 + \frac{1}{2}\|\mathbf{a}_i\|_2^2 \tag{3.63}$$

where $\mathbf{a}_i = [x_i \quad y_i \quad z_i]^{\mathrm{T}}$.

The sensitivity of the measurements is given by the gradient of (3.63) as

$$\frac{\partial \mathbf{a}_i\mathbf{x}}{\partial \mathbf{x}} + c\frac{\partial \tau_i \|\mathbf{x}\|_2}{\partial \mathbf{x}} = -\frac{1}{2}c^2\frac{\partial \tau_i^2}{\partial \mathbf{x}} \tag{3.64}$$

since the last term in (3.63) is a constant. This can be written as

$$\mathbf{a}_i^{\mathrm{T}} + c\left(\tau_i \frac{\mathbf{x}}{\|\mathbf{x}\|_2} + \|\mathbf{x}\|_2 \frac{\partial \tau_i}{\partial \mathbf{x}}\right) = -c^2\tau_i \frac{\partial \tau_i}{\partial \mathbf{x}} \tag{3.65}$$

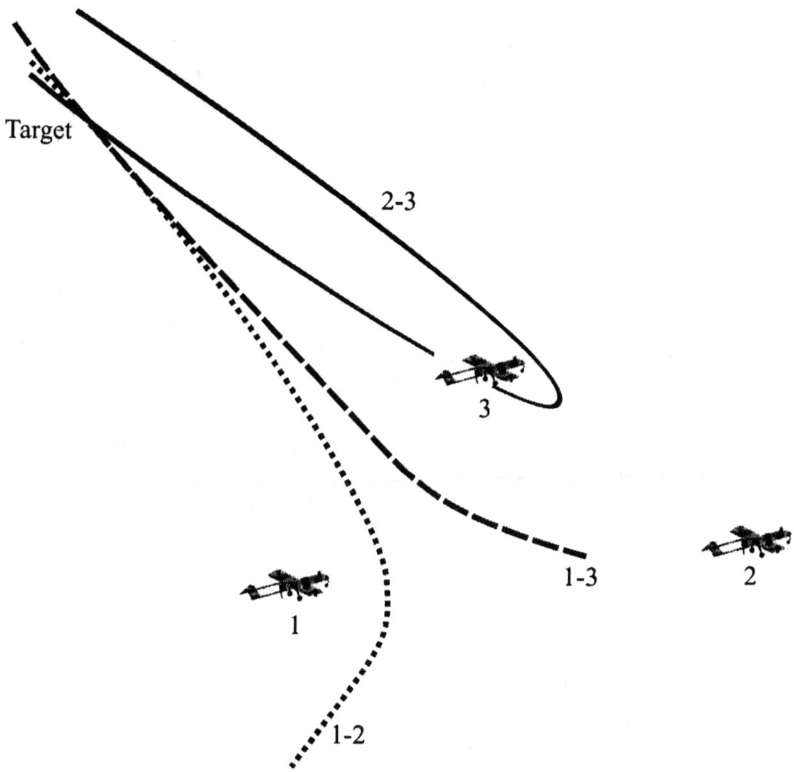

Figure 3.14 An illustration of the errors caused by the GDOP effect with TDOA processing.

The solution to this partial differential equation is given by [17]

$$\frac{\partial \tau_i}{\partial \mathbf{x}} = \frac{1}{c}\left(\frac{\mathbf{x} - \mathbf{a}_i^{\mathrm{T}}}{\left\| \mathbf{x} - \mathbf{a}_i^{\mathrm{T}} \right\|_2} - \frac{\mathbf{x}}{\left\| \mathbf{x} \right\|_2} \right) \tag{3.66}$$

A gradient vector with entries from (3.66) is defined as

$$\mathbf{H} = \begin{bmatrix} \frac{\partial \tau_1}{\partial \mathbf{x}} & \frac{\partial \tau_2}{\partial \mathbf{x}} & \cdots & \frac{\partial \tau_N}{\partial \mathbf{x}} \end{bmatrix}^{\mathrm{T}} \tag{3.67}$$

Let g denote the GDOP. Then g is given by [18]

$$g = \sqrt{\mathrm{tr}(\mathbf{H}^{\mathrm{T}}\mathbf{H})^{-1}} \tag{3.68}$$

If the variances of the measurements are not the same at all sensors, then this is taken into consideration by weighting matrix **W**. In this case,

$$\mathbf{Q} = \begin{bmatrix} \sigma_1^2 & 0 & \cdots & 0 \\ 0 & \sigma_2^2 & \cdots & 0 \\ \vdots & \vdots & \ddots & \vdots \\ 0 & 0 & \cdots & \sigma_M^2 \end{bmatrix} \quad (m^2) \tag{3.69}$$

Then the GDOP becomes

$$g = \sqrt{\mathrm{tr}(\mathbf{H}^{\mathrm{T}}\mathbf{W}\mathbf{H})^{-1}} \tag{3.70}$$

where $\mathbf{W} = \sigma_b^2 \mathbf{Q}^{-1}$ and σ_b^2 is a selected normalizing variance. The spherical uncertainty is then given by

$$\sigma_{3D} = \sigma_b g \tag{3.71}$$

In [17], an example of the effects of GDOP is illustrated. The sensors consist of a near planar circular array of a certain radius; the radius is referred to as the baseline. The results of GDOP-induced error with this array are illustrated in

Figure 3.15. At a range of only two baselines the accuracy degrades by a factor of 20 to 60.

3.1.9 Bias Effects of TDOA PF Estimation

Just as LOB measurement bias errors cause errors in estimating a PF with triangulation, errors are caused by measurement bias in TDOA calculations as well. An analysis and simulation was presented by Koorapaty, Grubeck, and Cedervall [19] to examine such effects when TDOA and TOA measurements were used to estimate the location of 911 callers. Their results are based on the 911 emergency call requirements and the specific results are couched in the vernacular of this requirement. Nevertheless, they illustrate the effects of bias on TDOA PF estimation.

The system equations are given by

$$r_i = f_i(\mathbf{x}) + b_i + n_i, \qquad i = 1, 2, \ldots, M \tag{3.72}$$

where b_i represents the bias in measurement I and $\mathbf{x} = [x_\text{T} \quad y_\text{T} \quad z_\text{T}]^\text{T}$ is the location of the target. In general, $f_i(\mathbf{x})$ is nonlinear, so a Taylor expansion is determined and only the linear terms are kept. In matrix form, this becomes

$$\mathbf{r} = \mathbf{f}(\mathbf{x}) + \mathbf{b} + \mathbf{n} \tag{3.73}$$

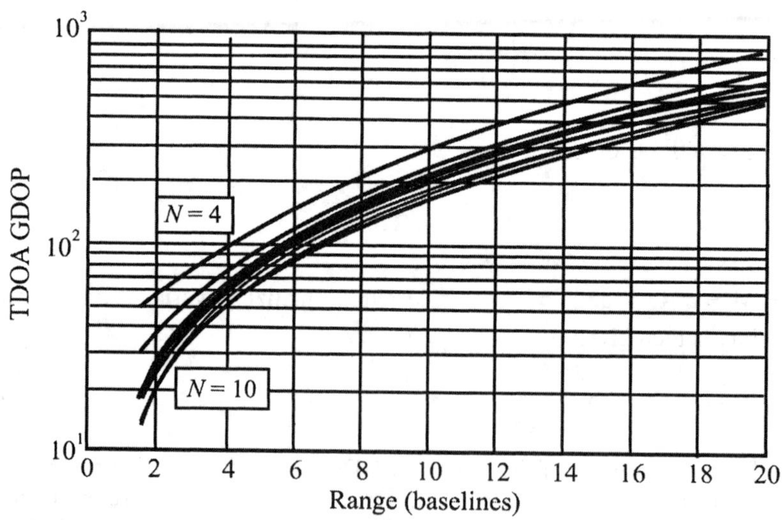

Figure 3.15 TDOA GDOP for N = 4 to 10 sensor elements. (*Source:* [17], © 1999 IEEE. Reprinted with permission.)

The least-squared estimator finds that $\hat{\mathbf{x}}$ that minimizes

$$Q(\mathbf{x}) = [\mathbf{r} - \mathbf{f}(\mathbf{x}) - \mathbf{b}]^{\mathrm{T}} \mathbf{N}^{-1} [\mathbf{r} - \mathbf{f}(\mathbf{x}) - \mathbf{b}] \tag{3.74}$$

When the elements of **n** are jointly Gaussian with covariance matrix **N**, then this process is also the maximum likelihood estimator.

It is assumed that the biases were unknown but nonrandom parameters, constant over the measurement interval. It was also assumed that the target remained fixed over the measurement interval.

The Taylor series approximation around reference point $\mathbf{x}_0$ is given by

$$\mathbf{f}(\mathbf{x}) \approx \mathbf{f}(\mathbf{x}_0) + \mathbf{G}_0(\mathbf{x} - \mathbf{x}_0) \tag{3.75}$$

where

$$\mathbf{G}_0 = \begin{bmatrix} \left.\dfrac{\partial f_1(\mathbf{x})}{\partial x_1}\right|_{\mathbf{x}=\mathbf{x}_0} & \cdots & \left.\dfrac{\partial f_1(\mathbf{x})}{\partial x_k}\right|_{\mathbf{x}=\mathbf{x}_0} \\ \vdots & \ddots & \vdots \\ \left.\dfrac{\partial f_M(\mathbf{x})}{\partial x_1}\right|_{\mathbf{x}=\mathbf{x}_0} & \cdots & \left.\dfrac{\partial f_M(\mathbf{x})}{\partial x_k}\right|_{\mathbf{x}=\mathbf{x}_0} \end{bmatrix} \tag{3.76}$$

Therefore, from (3.74) and (3.75),

$$Q(\mathbf{x}) = \left(\mathbf{r}_1 - \mathbf{G}_0\mathbf{x} - \mathbf{b}\right)^{\mathrm{T}} \mathbf{N}^{-1} \left(\mathbf{r}_1 - \mathbf{G}_0\mathbf{x} - \mathbf{b}\right) \tag{3.77}$$

is the function to be minimized where

$$\mathbf{r}_1 = \mathbf{r} - \mathbf{f}(\mathbf{x}_0) + \mathbf{G}_0\mathbf{x}_0 \tag{3.78}$$

To ascertain where the minimum lies, the gradient vector is calculated,

$$\nabla_{\mathbf{x}} Q(\mathbf{x}) = \begin{bmatrix} \dfrac{\partial Q}{\partial x_1} & \dfrac{\partial Q}{\partial x_2} & \cdots & \dfrac{\partial Q}{\partial x_k} \end{bmatrix}^{\mathrm{T}} \tag{3.79}$$

and that value of **x** such that $\nabla_{\mathbf{x}} Q(\mathbf{x}) = \mathbf{0}$ is determined.

If it is assumed that the biases $\mathbf{b}$ are not a function of $\mathbf{x}$ in the region around the sensor yielding $\nabla_x \mathbf{b} = 0$ and that $\mathbf{G}_0^T \mathbf{N}^{-1} \mathbf{G}_0$ is nonsingular, then that value of $\mathbf{x}$ that yields the minimum $Q(\mathbf{x})$ is given by

$$\hat{\mathbf{x}} = (\mathbf{G}_0^T \mathbf{N}^{-1} \mathbf{G}_0)^{-1} \mathbf{G}_0^T \mathbf{N}^{-1} \mathbf{r}_1 - (\mathbf{G}_0^T \mathbf{N}^{-1} \mathbf{G}_0)^{-1} \mathbf{G}_0^T \mathbf{N}^{-1} \mathbf{b} \tag{3.80}$$

If the biases are known then (3.80) can be used to determine the best estimate of the target location. If the biases are not known, then they are assumed to be zero and, using (3.78),

$$\hat{\mathbf{x}} = (\mathbf{G}_0^T \mathbf{N}^{-1} \mathbf{G}_0)^{-1} \mathbf{G}_0^T \mathbf{N}^{-1} [\mathbf{r} - \mathbf{f}(\mathbf{x}_0)] + \mathbf{x}_0 \tag{3.81}$$

and, using (3.73),

$$\hat{\mathbf{x}} = (\mathbf{G}_0^T \mathbf{N}^{-1} \mathbf{G}_0)^{-1} \mathbf{G}_0^T \mathbf{N}^{-1} [\mathbf{f}(\mathbf{x}) - \mathbf{f}(\mathbf{x}_0) - \mathbf{G}_0(\mathbf{x} - \mathbf{x}_0) + \mathbf{b} + \mathbf{n})] + \mathbf{x} \tag{3.82}$$

The indication of how the bias affects the PF estimate is given by the CEP. Assuming that the measurement errors are Gaussian, a simulation was performed using the geometry illustrated in Figure 3.16. The base stations are indicated by the black dots. The CEP was calculated at 100 equally spaced points within the square and the average value over the area was used as the figure of merit. The standard deviation of the measurements at the base stations was assumed to be 50m. The mean values of bias were assumed to be 0, 50, 100, 150, and 200 meters as one set and 0, 25, 50, 75, and 100 meters as another set. One hundred evenly spaced points within the square were selected and the mean CEP computed for each of them assuming the bias values given. A comparison of the results is shown in Figure 3.17. This cumulative distribution function shows the fraction of cells with CEPs less than the percentage shown on the abscissa, which is the percentage of the 125m CEP. The rest of the CEPS were larger than 125m. Clearly, the performance with smaller biases produced significantly more CEPs within the 125m limit imposed by the FCC.

3.1.10 Effects of Movement on TDOA PF Estimation

When there is relative motion between the target and sensors, which would normally be the case for airborne sensors, then errors in TDOA measurements arise. Chan and Ho call this effect the *scale difference of arrival* (SDOA) [20] and provide a method of estimating its effects. The TDOA error induced by this motion is proportional to the SDOA and the data record length.

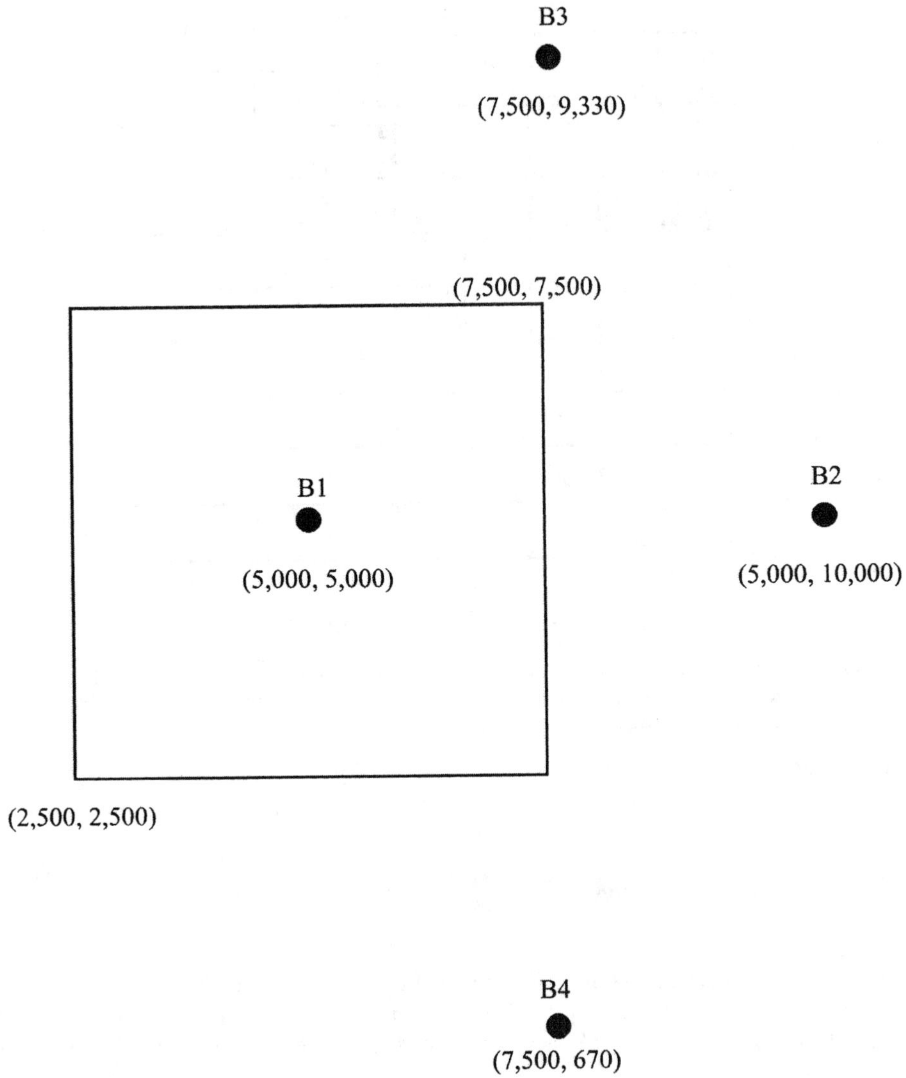

Figure 3.16 Geometry to simulate the effects of measurement bias on TDOA PF estimation in a cellular phone system. (*Source:* [19], © 1998 IEEE. Reprinted with permission.)

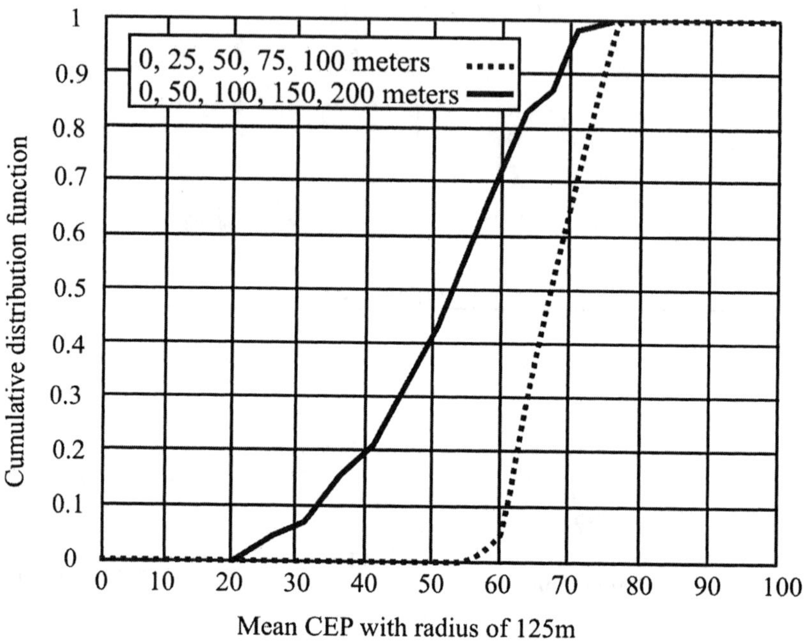

Figure 3.17 Cumulative distribution function of the CEP considering the effects of bias. The abscissa is the percentage within 125m. The others were larger than 125m. The mean measurement variance = 2,500m^2. (*Source:* [19], © 1998 IEEE. Reprinted with permission.)

The signals received at the two sensors are given by

$$r_1(t) = s(t) + \phi(t) \tag{3.83}$$

$$r_2(t) = s\left(\frac{t+D}{a}\right) + \varphi(t) \tag{3.84}$$

where $s(t)$ is the transmitted signal, $\phi(t)$ and $\varphi(t)$ are noise sources independent of each other and of the signal, D is the TDOA sought, and a is the SDOA. The joint estimate of the TDOA and SDOA can be obtained from the CAF between the two signals given by

$$\text{CAF} = \int_0^T r_1(t) r_2(\alpha t - \tau) dt \tag{3.85}$$

by finding the values of α and τ that maximize (3.85). Note that when there is no noise present,

$$r_2(\alpha t-\tau)=s\left(\frac{\alpha t-\tau+D}{a}\right) \tag{3.86}$$

so that (3.85) is maximized when $\alpha = a$ and $\tau = D$.

The TDOA estimate from maximizing (3.85) *neglecting SDOA* is given by

$$\tau=(1-a)t+D \tag{3.87}$$

so the bias error is

$$t-D=(1-a)t \tag{3.88}$$

Averaging (3.88) from 0 to N, where N is the number of data samples, yields

$$\varepsilon(a,N)=\frac{1}{N}\int_0^N(1-a)t\,dt=\frac{(1-a)N}{2} \tag{3.89}$$

The effects of the data record length on the TDOA bias are illustrated in Figure 3.18. Here the bias, neglecting the SDAO when there is no noise, with a = 1.001 and D = 0, is plotted versus the record length. Clearly from (3.89) the bias is a linear function of the record length N.

The reference contains a fast algorithm for computing the TDOA MSE and SDOA MSE from the CAF. The results for one case are illustrated in Figures 3.19 and 3.20. Figure 3.19 illustrates the MSE estimation error on the TDOA versus the SNR given by

$$\nu=10\log\frac{\sigma_s^2}{\sigma_\phi^2} \tag{3.90}$$

where $\sigma_\phi^2=\sigma_\varphi^2$ and band-limited white noise is assumed. Likewise, Figure 3.20 shows the MSE estimation error of the SDOA versus SNR. At low SNR the effects on the TDOA MSE are significantly more pronounced than those on the SDOA. At ν = 0 dB, the SDOA MSE is 45 dB less than that for the TDOA.

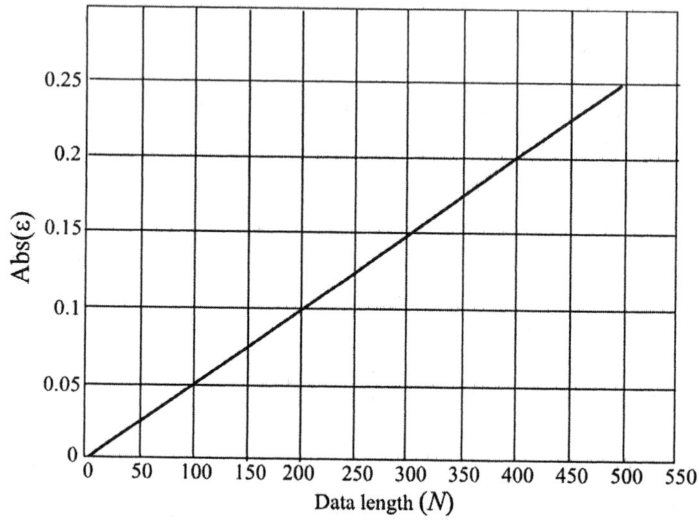

Figure 3.18 TDOA bias error caused by data record length, ignoring the SDOA. (*Source:* [20], © 2003 IEEE. Reprinted with permission.)

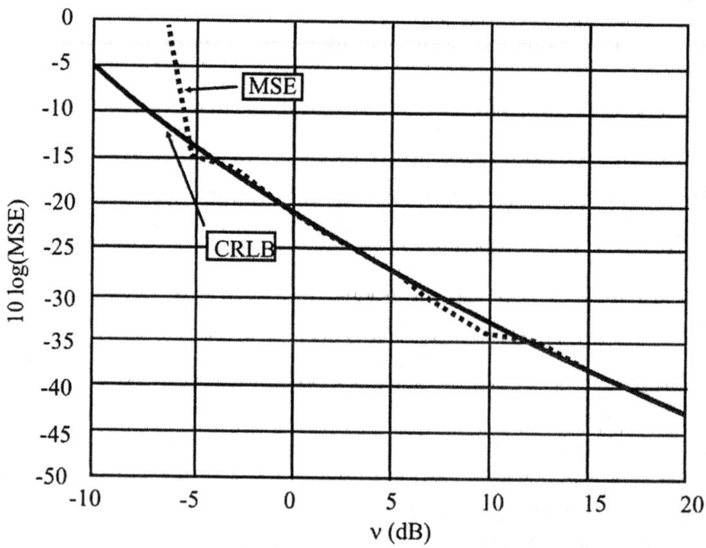

Figure 3.19 TDOA bias error caused by motion. (*Source*: [20], © 2003 IEEE. Reprinted with permission.)

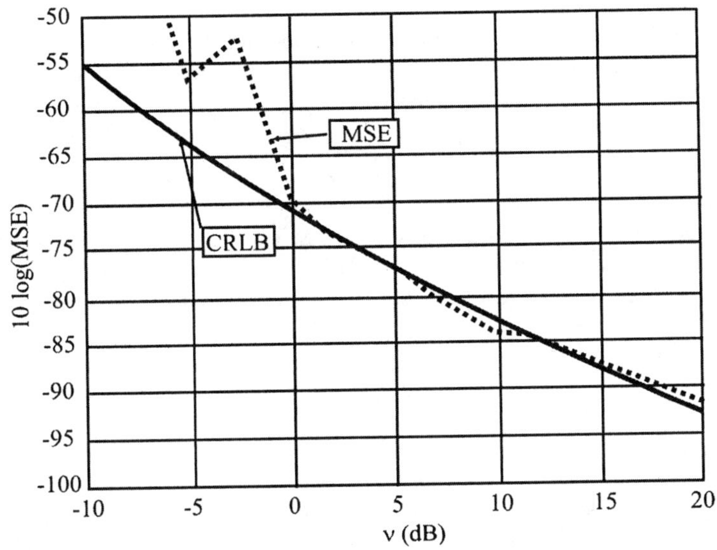

Figure 3.20 SDOA error caused by relative motion. (*Source:* [20], © 2003 IEEE. Reprinted with permission.)

3.2 Differential Doppler

Utilization of frequency difference information at two or more sensors for PF calculation is presented in this section. This follows that in [1] closely, and is included here for completeness.

3.2.1 Introduction

The signals emitted by a moving object exhibit an effect called Doppler shift which manifests itself in a frequency difference depending on the direction of travel of the moving object relative to a receiver. When the frequency of a signal is measured at two (or more) sensor sites, the frequency difference between the signals at the two sensors can be computed. This frequency difference, called DD, is a measurement that can be used for target geolocation estimation. To utilize the DD effect, only one object needs to be moving—one or more sensors or the target. The higher the velocity the greater the frequency difference, which makes the DD easier to measure accurately.

3.2.2 Differential Doppler

The DD, where the two sensors and/or target velocities are much smaller than the speed of light, is given by

$$\dot{\tau} = \frac{v_2}{c} f_0 - \frac{v_1}{c} f_0 = \frac{f_0}{\lambda}(v_2 - v_1) \tag{3.91}$$

where v_i represents the instantaneous velocity of the sensors relative to the transmitter in the radial direction and f_0 is the frequency of the transmitted signal. The DD is frequently denoted by $\dot{\tau}$ because it represents the difference in two velocities, which are, in turn, the time derivatives of the distances, to wit,

$$\dot{\tau} = \frac{f_0}{c}\left(\frac{dr_2}{dt} - \frac{dr_1}{dt}\right) \tag{3.92}$$

where r_i is the range between the transmitter and the sensor, as shown in Figure 3.1. Thus, dr_i/dt is the rate of change of the range in the radial direction. The distances are given by

$$r_i = \sqrt{(x_{\mathrm{T}} - x_i)^2 + y_{\mathrm{T}}^2} \tag{3.93}$$

yielding

$$\frac{dr_i}{dt} = \frac{d\left[(x_{\mathrm{T}} - x_i)^2 + y_{\mathrm{T}}^2\right]^{1/2}}{dt} \tag{3.94}$$

$$= \frac{(x_{\mathrm{T}} - x_i)}{\left[(x_{\mathrm{T}} - x_i)^2 + y_{\mathrm{T}}^2\right]^{1/2}} \frac{dx_i}{dt}, \qquad i = 1, 2 \tag{3.95}$$

where it is assumed that the aircraft are flying at the same speed parallel to the x-axis so that $dy_i/dt = 0$. Denoting $v = v_1 = dx_1/dt = v_2 = dx_2/dt$, then

$$\dot{\tau} = \frac{f_0 v}{c}\left\{\frac{(x_{\mathrm{T}} - x_1)}{\left[(x_{\mathrm{T}} - x_1)^{1/2} + y_{\mathrm{T}}^2\right]^{1/2}} - \frac{(x_{\mathrm{T}} - x_2)}{\left[(x_{\mathrm{T}} - x_2)^{1/2} + y_{\mathrm{T}}^2\right]^{1/2}}\right\} \tag{3.96}$$

The surface formed by this expression for the example here is shown in Figure 3.21. The DD curves are complex quadratic functions, not simple hyperbolas as was the case for TDOA processing. Viewed from the top they look as seen in Figure 3.22. The intersection of these curves for the appropriate value of $\dot{\tau}$ will yield candidate geolocations for the emitter, but with only two sensors there is a left-right ambiguity as seen in Figures 3.23 and 3.24. As with TDOA, adding a third sensor can resolve the ambiguity and produce a unique solution, as shown in Figure 3.25.

3.2.3 DD Measurement Accuracy

Just as for TDOA, at extreme ranges from the baselines the iso-Doppler curves become nearly parallel, as illustrated in Figure 3.26. For accurate PF estimates it is desirable for the iso-Doppler lines to cross at right angles. When they are nearly parallel as in Figure 3.26, poor PF estimates result.

The Cramer-Rao bound for estimating $\dot{\tau}$ is given by

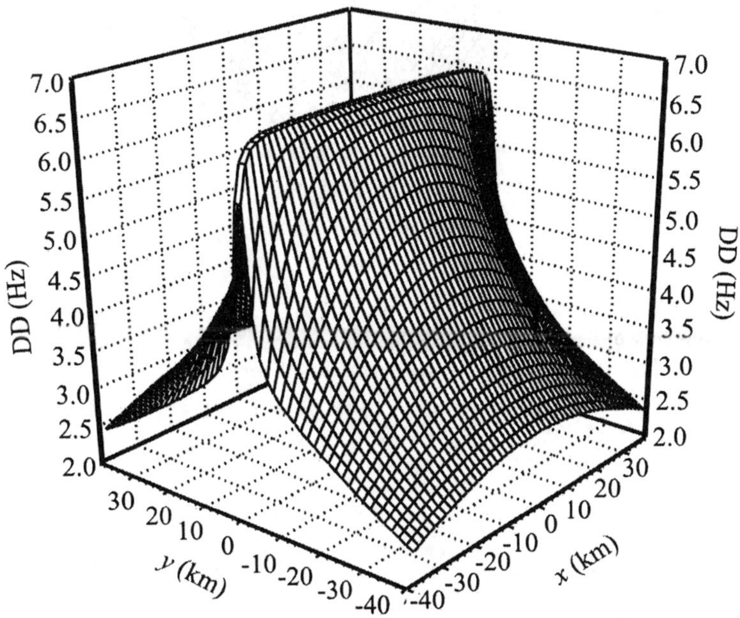

Figure 3.21 Surface of the DD for the example. In this case v = 10 m/sec and f = 100 MHz (λ = 3m). (*Source:* [1], © 2002 Horizon House. Reprinted with permission.)

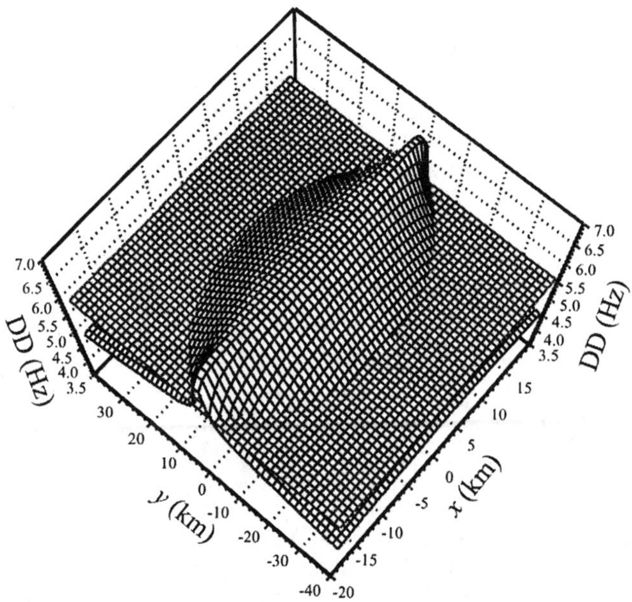

Figure 3.22 Top view of the $\dot{\tau}$ contour. The shape resembles an ellipse, but actually it is a complex quadratic. (*Source:* [1], © 2002 Horizon House. Reprinted with permission.)

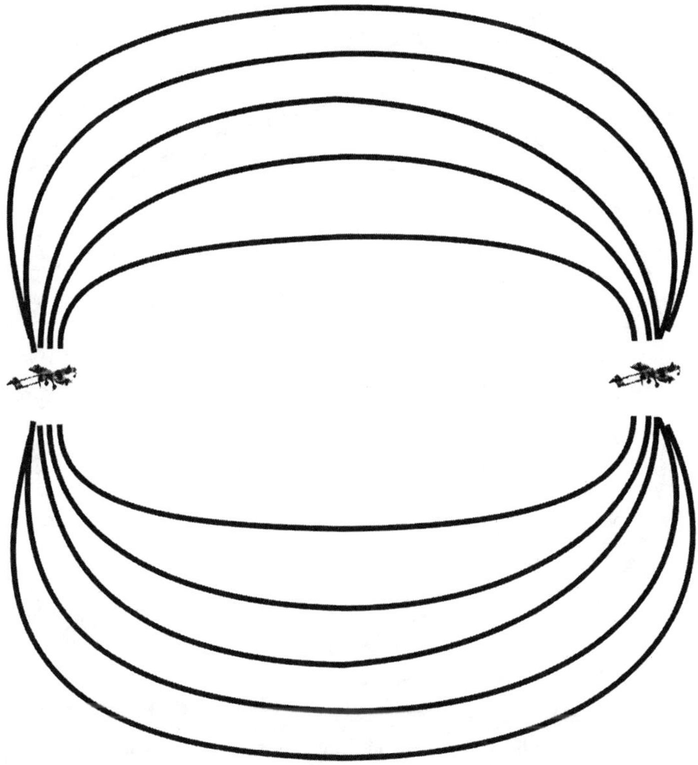

Figure 3.23 Lines of constant $\dot{\tau}$ are complex quadratic curves.

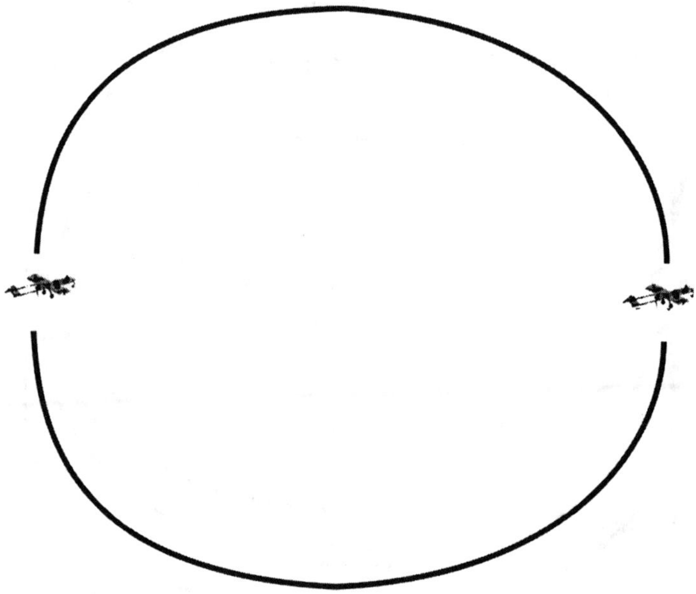

Figure 3.24 DD contours corresponding to an emitting target. The contours have a left–right ambiguity.

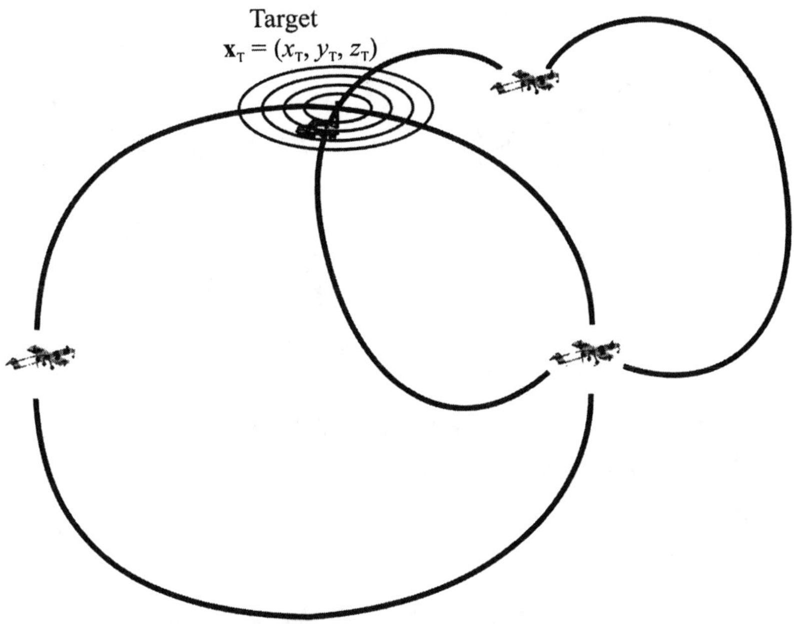

Figure 3.25 Iso-Doppler contours for three sensors can uniquely locate the target.

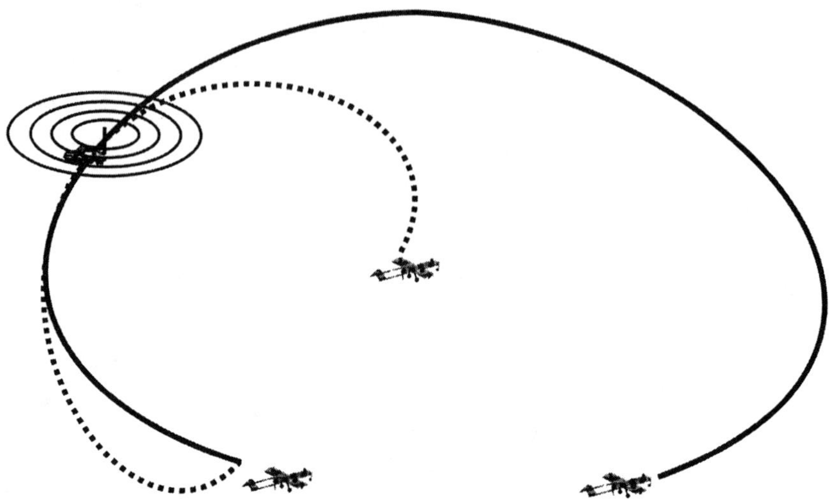

Figure 3.26 Iso-Doppler contours for a distant target are nearly parallel.

$$\sigma_{\dot{\tau}} = \frac{1}{T_e}\frac{1}{\sqrt{BT\gamma}} \tag{3.97}$$

where B, T, and γ are as above, $\sigma_{\dot{\tau}}$ is the standard deviation of the $\dot{\tau}$ measurement, and T_e is the RMS integration time given by

$$T_e = 2\pi\left[\frac{\int_{-\infty}^{\infty} t^2 |u(t)|^2\, dt}{\int_{-\infty}^{\infty} |u(t)|^2\, dt}\right]^{1/2} \tag{3.98}$$

Where $u(t)$ is the probability density function of the integration time. Again, for example, if the actual integration time is T and the pdf for the integration time has sharp edges, then $T_e = 2\pi T/\sqrt{3}$. Equation (3.97) is plotted in Figure 3.27 for a few values of integration time versus SNRs. As indicated in Figure 3.23, the DD values for representative values of parameters are in the single-to-double digit hertz range. The standard deviation values indicated in Figure 3.27 for low SNRs

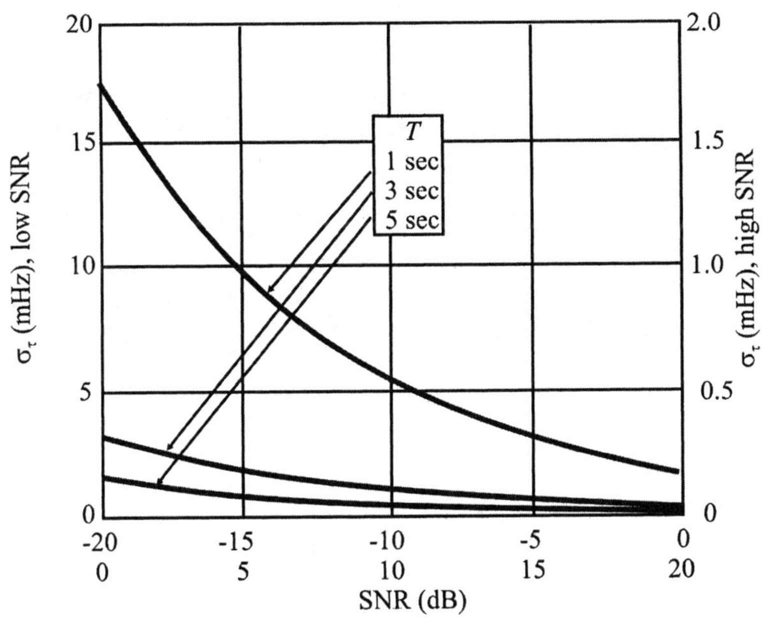

Figure 3.27 The CRLB for DD processing versus SNRs assuming a uniform distribution for T_e and $B = 25$ kHz.

are only about 1/100 those values, so the possible accuracy at these integration times may be suspect. Higher SNRs produce better fixes.

3.2.4 Maximum Likelihood DD Algorithms

A DD MLE algorithm similar to the LOB MLE algorithm described in Section 2.6 has been developed by Levanon [21]. A PF system that uses DD measurements avoids the ambiguity problem inherent in interferometer PF systems. The antennas on the same platform can be separated by more than $\lambda/2$ and effective measurements can be obtained. On an airborne sensor platform, however, relatively dense measurements are required. For an airborne sensor, the Doppler measurements must be made often enough so that at least one measurement is taken during each interval defined by the wavelength of the signal.

An interferometer, on the other hand, must have the sensor receivers spaced less than $\lambda/2$ to avoid ambiguous results.

The DD is determined by utilizing many measurements of the phase, and, therefore, range differences along the path. Each time the phase jumps by 2π, one more wavelength in range has been traversed. These jumps are counted and used as a measurement of the differential frequency.

3.2.4.1 DD MLE Algorithm

The unknowns in the geometry shown in Figure 2.36, repeated here as Figure 3.28, are y_c, the center of the observation interval, and x_T, the location of the target (it is assumed for now that $y_T = 0$). Therefore, the vector of unknowns is

$$\mathbf{x} = \begin{bmatrix} x_T & y_c \end{bmatrix}^T \tag{3.99}$$

Denote the noiseless relationship between the unknowns and the measurement during the kth observation as $\tilde{z}_k(x_T, y_c)$. For the case of the Doppler measurements here,

$$\tilde{z}_k(x_T, y_c) = R_{1k} - R_{2k} - (R_{10} - R_{20}) \tag{3.100}$$

where, for $k = 0, \pm1, \pm2, \ldots, \pm M$, $N = 2M + 1$,

$$R_{1k} = \sqrt{\left(y_c + \frac{b}{2} + k\Delta L\right)^2 + x_T^2}$$

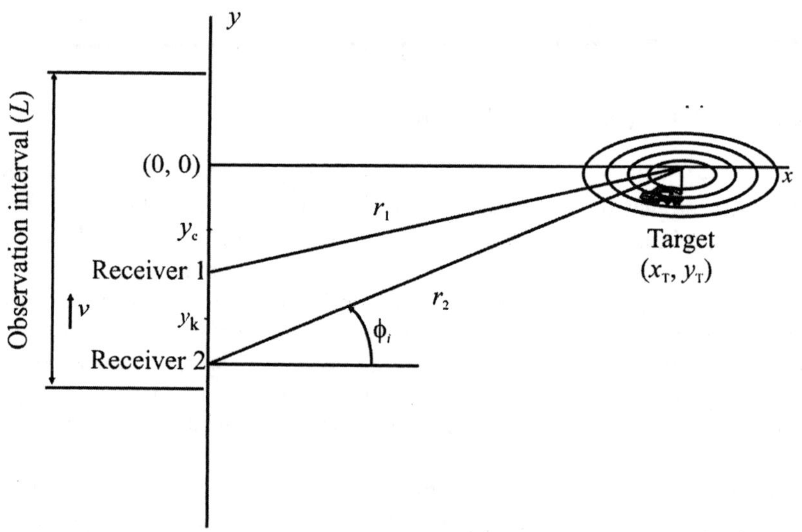

Figure 3.28 MLE geometry. (*Source:* [22], © 1989 IEE. Reprinted with permission.)

$$R_{2k} = \sqrt{\left(y_c - \frac{b}{2} + k\Delta L\right)^2 + x_T^2}$$

$\Delta L =$ distance traveled between measurements

With these values of k, the reference is at the center of the observation interval L. The $2M + 1$ noisy measurements, denoted by z_k, corrupted by samples of the Gaussian noise process denoted by n_k, are given by

$$z_k(x_T, y_c) = \tilde{z}_k(x_T, y_c) + n_k, \qquad k = 0, \pm 1, \pm 2, \ldots, \pm M \tag{3.101}$$

The final covariance matrix for the variables gives the PF error. It is given by

$$\text{cov}(\mathbf{x}) = \sigma_n^2 (\mathbf{G}^T \mathbf{G})^{-1} \tag{3.102}$$

The error in the 0th range difference measurement ε must be accommodated in the noiseless measurement (3.100), changing it to

$$\tilde{z}_k(x_T, y_c) = R_{1k} - R_{2k} - (R_{10} - R_{20}) + \varepsilon \tag{3.103}$$

where ε is an unknown and therefore must be estimated along with x_T and y_c. Thus, the unknown vector to be estimated is

$$\mathbf{x} = [x_T \quad y_c \quad \varepsilon]^T \tag{3.104}$$

With two simplifying assumptions that $y_c = 0$ and $M\Delta L \ll x_T$, then

$$\partial_{kx} = \frac{\partial \tilde{z}_k(x_T, y_c)}{\partial x_T} \tag{3.105}$$

$$\approx -\frac{bk\Delta L}{x_T^2} \tag{3.106}$$

and

$$\partial_{ky} = \frac{\partial \tilde{z}_k(x_T, y_c)}{\partial y_c} \tag{3.107}$$

$$\approx -\frac{3b(k\Delta L)^2}{2x_\mathrm{T}^3} \tag{3.108}$$

The **G** matrix now is

$$\mathbf{G} = \begin{bmatrix} \partial_{Mx} \cdots \partial_{kx} \cdots \partial_{Mx} \\ \partial_{-My} \cdots \partial_{ky} \cdots \partial_{My} \\ 1 \cdots 1 \cdots 1 \end{bmatrix}^\mathrm{T}$$

which, using (3.105) and (3.107), generates

$$\mathbf{G}^\mathrm{T}\mathbf{G} = \begin{bmatrix} \partial_{-Mx} \cdots \partial_{kx} \cdots \partial_{Mx} \\ \partial_{-My} \cdots \partial_{ky} \cdots \partial_{My} \\ 1 \cdots 1 \cdots 1 \end{bmatrix} \begin{bmatrix} \partial_{-Mx} & \partial_{-My} & 1 \\ \vdots & \vdots & \vdots \\ \partial_{ky} & \partial_{ky} & 1 \\ \vdots & \vdots & \vdots \\ \partial_{Mx} & \partial_{My} & 1 \end{bmatrix}$$

$$= \begin{bmatrix} \sum_{k=-M}^{M} (\partial_{kx})^2 & \sum_{k=-M}^{M} \partial_{kx}\partial_{ky} & \sum_{k=-M}^{M} \partial_{kx} \\ \sum_{k=-M}^{M} \partial_{kx}\partial_{ky} & \sum_{k=-M}^{M} (\partial_{ky})^2 & \sum_{k=-M}^{M} \partial_{ky} \\ \sum_{k=-M}^{M} \partial_{kx} & \sum_{k=-M}^{M} \partial_{ky} & \sum_{k=-M}^{M} 1 \end{bmatrix}$$

Assuming that ΔL is small enough that it can be approximated by a differential, then converting these sums to integrals, dividing (outside the sum) by dL, and multiplying (inside the sum) by dL yields

$$\mathbf{G}^\mathrm{T}\mathbf{G} = \begin{bmatrix} \dfrac{Nb^2L^2}{12x_\mathrm{T}^4} & 0 & 0 \\ 0 & \dfrac{9Nb^2L^4}{320x_\mathrm{T}^6} & \dfrac{NbL^2}{8x_\mathrm{T}^3} \\ 0 & \dfrac{NbL^2}{8x_\mathrm{T}^3} & N \end{bmatrix} \tag{3.109}$$

and

$$(\mathbf{G}^{\mathrm{T}}\mathbf{G})^{-1} = \begin{bmatrix} \dfrac{12x_{\mathrm{T}}^4}{Nb^2L^2} & 0 & 0 \\ 0 & \dfrac{80x_{\mathrm{T}}^6}{Nb^2L^4} & -\dfrac{10x_{\mathrm{T}}^3}{NbL^2} \\ 0 & -\dfrac{10x_{\mathrm{T}}^3}{NbL^2} & \dfrac{9}{4N} \end{bmatrix} \tag{3.110}$$

Now from (3.102) and (3.110),

$$\sigma_x \approx \sqrt{\frac{12}{N}}\frac{\sigma_{\Delta R}}{b}\frac{x_{\mathrm{T}}^2}{L} \tag{3.111}$$

$$\sigma_y \approx \sqrt{\frac{80}{N}}\frac{\sigma_{\Delta R}}{b}\frac{x_{\mathrm{T}}^3}{L^2} \tag{3.112}$$

When $y_{\mathrm{c}} \neq 0$, (3.111) becomes more complicated while (3.112) remains the same. Thus,

$$\sigma_x \approx \sqrt{\frac{12}{N}}\frac{\sigma_{\Delta R}}{b}\frac{x_{\mathrm{T}}^2}{L}\sqrt{1+60\frac{y_c^2}{L^2}} \tag{3.113}$$

3.2.4.2 Performance

The DD results are portrayed graphically in Figure 3.29. These values of the parameters would correspond to a UAV platform that is prosecuting a fixed-frequency target that is transmitting a considerable length of time. A UAV flying at 100 kph would require about 1 minute to fly 1.6 km, so the target must be transmitting for at least that length of time.

Comparing Figure 3.29 with Figure 2.36 for the interferometric algorithm, it can be seen that the DD performance is better in the down-range estimation, while the interferometer is better in the cross-range results. The DD performance would be expected to be good, especially when $y_{\mathrm{c}} = 0$, because at that point the quadratic is intersecting with the x-axis at right angles, yielding the highest sensitivity possible from the DD measurements (see Figure 3.30). The fact that the variance is a minimum can be established by finding the derivative of (3.113) with respect to y_{c} and setting it to zero.

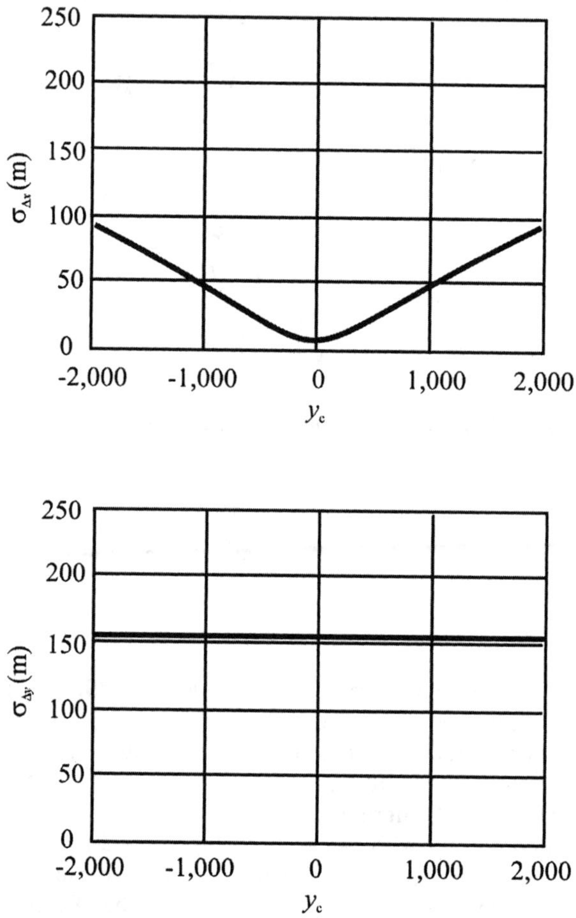

Figure 3.29 Differential Doppler results. For this example $N = 40$, $x_T = 10$ km, $L = 1.6$ km, and $\sigma_{\Delta R} =$ 0.0141m. (*Source:* [21], © 1989 IEE. Reprinted with permission.)

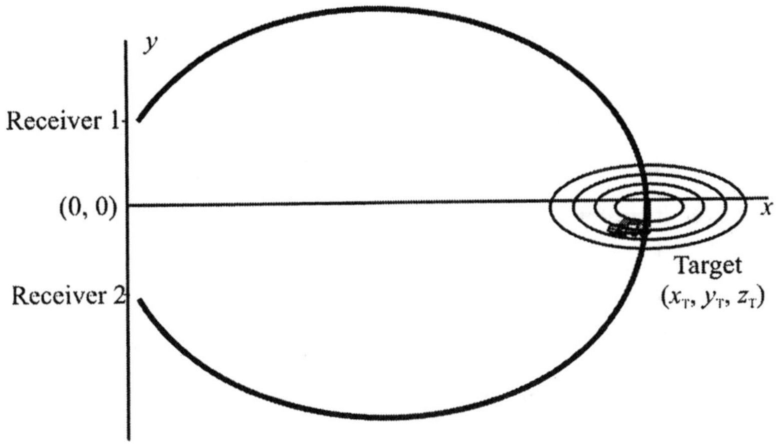

Figure 3.30 Intersection of the iso-Doppler contour with the x-axis when $y_c = 0$.

The highest sensitivity in the interferometric measurement also occurs at $y_c = 0$, as seen in Figure 3.31. At that point, the slope of the $\Delta\phi$ versus y_c curve is maximized because it is an inflection point. The difference in phase of the signals impinging on the two antennas is zero since the arrival azimuths are orthogonal to the y-axis. Again, the minimum variance point can be shown to be at $y_c = 0$ by taking the derivative of (2.125) with respect to y_c and setting it equal to zero.

3.2.5 Cross Ambiguity Function

TDOA and DD calculations can be combined to take advantage of the benefits of each, as well as possibly increase the accuracy of the PF estimates. This is typically accomplished with the CAF. This computation simultaneously yields the τ and $\dot{\tau}$ for two sensors, and, just as for TDOA isochrones and iso-Doppler curves, the CAF must be computed for each pair of sensors in order to yield sufficient information to calculate an estimate. The CAF is a generalization of the cross-correlation function and is given by

$$\text{CAF}(\tau, t) = \int_0^T s_1(t) s_2(t+\tau) e^{-j\omega t} dt \tag{3.114}$$

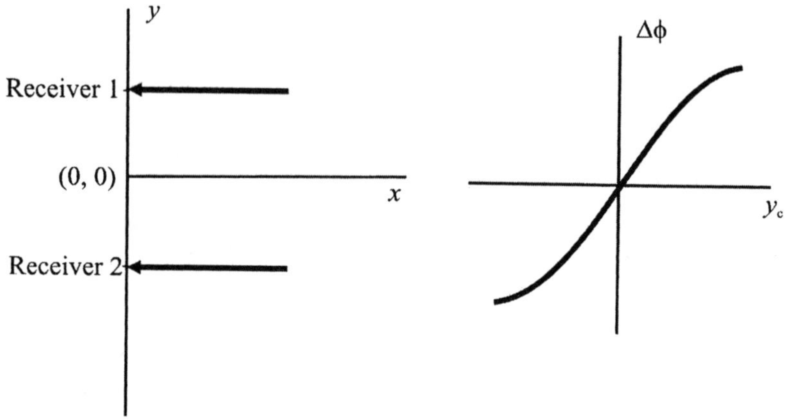

Figure 3.31 Least-squared LOB geometry when $y_c = 0$.

Under ideal conditions, a plot of the amplitude of the CAF versus τ and $\dot{\tau}$ displays a peak at the proper value of these variables.

As indicated in [1], the standard deviation of the measurements of the PF coordinates, calculated on the perpendicular bisector of the baseline shown in Figure 3.1, depends on these equations as follows:

$$\sigma_x = \frac{c\sigma_\tau \sqrt{\left(\frac{b}{2}\right)^2 + d^2}}{d} \tag{3.115}$$

and

$$\sigma_y = \frac{\lambda \sigma_{\dot{\tau}} \left[\left(\frac{b}{2}\right)^2 + d^2\right]^{3/2}}{vbd} \tag{3.116}$$

where λ is the wavelength, v is the velocity of the sensors, b is the baseline length, and d is the distance from the baseline.

As with TDOA and DD processing alone, as the target deviates from the perpendicular bisector, GDOP comes into play and the accuracy degrades.

3.2.6 Estimating the DD of a Sinusoid in Noise Using Phase Data

Otnes derived a method of estimating the DD of a sinusoidal signal in noise received at two or more platforms using regression techniques [22]. In other words, the DD is estimated as the slope of the line that is the linear regression of the DD data. This is a similar approach to that described in Section 3.1.8, which discussed a regression matching technique for estimating the TDOA derived by Piersol [6].

The signal transmitted by the target is given by

$$s(t) = Ae^{jw_0 t + \phi_0} \tag{3.117}$$

where ω_0 is the frequency and ϕ_0 is the phase offset. The signals received at two geographically dispersed sensors j and k are given by

$$s_j(t) = s(t) + n_j(t) \tag{3.118}$$

$$s_k(t) = s(t - \tau_{jk}) + n_k(t) \tag{3.119}$$

where it is assumed that the noises are uncorrelated to each other or to the transmitted signal. It is also assumed that there is relative motion between one or more sensors and the target. Normalizing the received signals so the time delay can be ignored and using relative phase delays so $\phi_0 = 0$ results in the received signals given by

$$s_j(t) = A_j e^{j(\omega_j t + \phi_j)} + n_j(t) \tag{3.120}$$

$$s_k(t) = A_k e^{j(\omega_k t + \phi_k)} + n_k(t) \tag{3.121}$$

The *mixing product* [7] of these two signals is given by

$$p_{jk}(t) = s_j(t) s_k^*(t) \tag{3.122}$$

Because of the assumed independence of the noise from the signals,

$$\mathcal{E}\{p_{jk}(t)\} = A_j A_k^* e^{j(\omega_{jk} t + \phi_{jk})} \tag{3.123}$$

where $\omega_{jk} = \omega_j - \omega_k$ and $\phi_{jk} = \phi_j - \phi_k$.

Using (3.120) and (3.121) in (3.122) gives

$$p_{jk}(t) = A_j A_k^* e^{j(\omega_{jk}t + \phi_{jk})} + n_{jk}(t) \tag{3.124}$$

$$= Ae^{j\omega_{jk}t} + n_{jk}(t) \tag{3.125}$$

where $A = A_j A_k^* e^{j\phi_{jk}}$ and $n_{jk}(t)$ is the noise term consisting of the remaining products resulting from (3.122) incorporating (3.120) and (3.121).

Ignoring noise, the phase of the mixing product after the phase has been unwrapped is given by

$$y(t) = \tan^{-1}\frac{\text{Im}[p_{jk}(t)]}{\text{Re}[p_{jk}(t)]} = \tan^{-1}\frac{|A|\sin(\omega_{jk}t + \phi_{jk})}{|A|\cos(\omega_{jk}t + \phi_{jk})} = \omega_{jk}t + \phi_{jk} \tag{3.126}$$

Thus, ignoring noise, the phase function is a linear function of time. With noise included and assuming a sufficiently high SNR, the phase function can be expressed as

$$y_{jk}(t) = \omega_{jk}t + \phi_{jk} + n(t) \tag{3.127}$$

and ω_{jk} and ϕ_{jk} can be determined using linear regression as discussed in Section 2.2, resulting in the MLSE estimate provided by (2.31).

Considering discrete time processing, during time interval i,

$$y_i = \omega_0 x_i + \phi_0 + n_i = \omega_0 iT + \phi_0 + n_i \tag{3.128}$$

because $x_i = iT$ with T the sampling interval. The optimum DD is given by

$$\omega_{jk}^* = \frac{\sum_{i=1}^{N} y_i(x_i - \bar{x})}{\sum_{i=1}^{N}(x_i - \bar{x})^2} \tag{3.129}$$

where $\bar{x}$ denotes the average value of x.

This expression can be reduced to

$$\omega_{jk}^* = \frac{\omega_0 TQ(N) + \sum_{i=1}^{N} n_i\left(i - \frac{N+1}{2}\right)}{TQ(N)} \tag{3.130}$$

where

$$Q(N) = \frac{2N(N-1)(N+1)}{24} \tag{3.131}$$

The corresponding variance is given by

$$\sigma^2_{\omega_{jk}} = \frac{\sum_{i=1}^{N} y_i(x_i - \overline{x}) - \frac{\left[\sum_{i=1}^{N} y_i(x_i - \overline{x})\right]^2}{\sum_{i=1}^{N}(x_i - \overline{x})^2}}{(N-2)\sum_{i=1}^{N}(x_i - \overline{x})^2} \tag{3.132}$$

The expected value of the variance for large N is given by

$$\mathcal{E}\{\sigma^2_{\omega_{jk}}\} \approx \frac{12\sigma_n^2}{T^2 N^3} \tag{3.133}$$

where σ_n^2 is the variance of the phase noise. Thus, the variance in the estimation of the DD decreases as the square of the sampling interval and the cube of the number of samples (N). This indicates that the accuracy in estimating the DD improves by these factors, illustrating the significant advantage of using a large number of samples.

Utilizing some assumptions, this method of estimating the DD will work when there is *amplitude modulation* (AM) applied to the sinusoidal carrier. In particular, when the DD value being sought is the fine value (as opposed to the coarse value which just gets close), then the amplitude of the modulation can be assumed to be equal to one and (3.120) and (3.121) still apply. This amplitude can be made to equal to one by clipping the signal, for example.

Because the parameter being sought is the differential frequency received at two sensor sites, this estimation method would not work well for signals with *frequency modulation* (FM) or *phase modulation* (PM).

3.2.7 Effects of Motion on DD PF Estimating

As discussed in Section 3.1.8, relative motion between the target and sensor can cause errors in the measurement of TDOA unless such motion can be taken into account. Even though it is necessary for there to be relative motion between the target and a receiver to be able to observe Doppler shift, assuming the transmitter to be stationary when it is in fact moving can lead to considerable error estimating the DD. It can be very difficult to take out the effects of a moving target on calculating the DD. An alternative is to attempt to ascertain that the target is moving or stationary, and, if moving, do not attempt to estimate the PF because it will contain substantial error.

Ullman and Geraniotis [23] developed two techniques for estimating when the transmitter is moving so that the data obtained can be ignored for PF computation purposes. Both techniques assume that the target position is confined to some surface and its velocity is constrained to a tangent plane. Also, both techniques form and test the hypotheses

$$H_0 : \mathbf{v}_e = \mathbf{0} \tag{3.134}$$

$$H_1 : \mathbf{v}_e \neq \mathbf{0} \tag{3.135}$$

where $\mathbf{v}_e$ is the actual velocity vector of the target.

The first technique is based on postulating that the transmitter is stationary. After several samples of TDOA/DD data are collected, the maximum likelihood stationary target position is found using these measurements and the probability of the measurements occurring given this position is determined. Let $\mathbf{q}(\mathbf{p}_e,\mathbf{v}_e)$ denote the TDOA/DD measurement vector. Let $\boldsymbol{\Sigma}$ denote the covariance matrix of measurements. The observed measurement vector, $\mathbf{q}_m$, consisting of noise corrupted measurements is assumed to be $\mathbf{q}_m \sim \mathcal{N}[\mathbf{q}(\mathbf{p}_e,\mathbf{v}_e), \boldsymbol{\Sigma}]$, where $\mathbf{p}_e$ is the actual position vector of the target. The maximum likelihood estimate is then given by

$$\tilde{\mathbf{p}}_e = \underset{p}{\operatorname{argmin}}[\mathbf{q}(\mathbf{p},\mathbf{0}) - \mathbf{q}_m]^{\mathrm{T}} \boldsymbol{\Sigma}^{-1} [\mathbf{q}(\mathbf{p},\mathbf{0}) - \mathbf{q}_m] \tag{3.136}$$

The second technique is based on a *likelihood ratio test* (LRT) (LRT-like actually, since it is not a true LRT) given by

$$\text{LRT} = \frac{\max\limits_{\mathbf{p}_e,\mathbf{v}_e} \Pr_1(\mathbf{q}_m | \mathbf{p}_e, \mathbf{v}_e)}{\max\limits_{\mathbf{p}_e} \Pr_0(\mathbf{q}_m | \mathbf{p}_e, \mathbf{v}_e = \mathbf{0})} \tag{3.137}$$

Both of these techniques yield similar results, although the specific results are highly scenario dependent, as one would expect, since the performance of the algorithms used for estimating the PF using TDOA and DD measurements are scenario dependent (due to GDOP, for example).

A comparison of the two techniques for specific examples illustrates that they produce similar results at detection of target motion. The test for stationarity is relatively simple compared to the LRT test, for which the computation complexity is substantially larger. Based on this result, the test for stationarity is the recommended approach in most situations.

An alternative to ignoring the data when target motion is detected is to estimate the velocity vector along with the target location. One such technique was developed by Rusu [24]. In that work, the TDOA and FDOA data are transformed into TOA and FOA data and closed-form expressions for the target location and velocity are used to determine the unknowns by using the implicit function theorem. The algebraic results are compared with simulation results.

Another technique for estimating the target velocity along with the location was developed by Ho and Xu [25]. Least squares minimizations are used to determine the unknowns. When the TDOA and FDOA errors are Gaussian, the technique is claimed to achieve the CRLB. The analytical results are compared favorably with the algebraic results.

3.3 Range Difference Methods

3.3.1 Introduction

The time of arrival of a signal at sensor i is given by

$$t_i = \frac{r_i}{c} \tag{3.138}$$

where r_i is the range from the target to sensor i. The corresponding TDOA of that signal arriving at two sensors is given by

$$(t_i - t_j) = \tau_{ij} = \frac{r_i - r_j}{c} \tag{3.139}$$

In the absence of errors due to inaccurate measurements or noise, the target must lie at the intersection of hyperboloids that are formed as the locus of constant RD as determined by (3.139). Therefore, finding this intersection is a method for establishing where the target lies.

Noise and measurement errors are always present, however, and normally these hyperboloids do not intersect at a single point but at several points. *Least-squared* (LS) techniques can then be used to determine the point where the total distance from that point to each of the hyperboloids is minimized.

3.3.2 Least-Squared Range Difference Methods

Let N represent the number of sensors and let Δ_{ij} represent the range difference between sensor i and the target and sensor j and the target. The location of sensor i is given by the coordinate vector $\mathbf{x}_i = [x_i \quad y_i \quad z_i]^\mathrm{T}$ and the location of the target is given by the vector $\mathbf{x}_\mathrm{T} = [x_\mathrm{T} \quad y_\mathrm{T} \quad z_\mathrm{T}]^\mathrm{T}$. The distance between the target and sensor i is given by

$$r_i = \left\| \mathbf{x}_i - \mathbf{x}_\mathrm{T} \right\|_2 \tag{3.140}$$

while the distance from the origin to point i is denoted by R_i.

Smith and Abel [26] formulated a method for minimizing the PF estimation error in an LS sense. They define the *equation error* as follows. Consider the geometry indicated in Figure 3.32 where, without loss of generality, sensor j is placed at the origin. Also without loss of generality, $j = 1$. Thus,

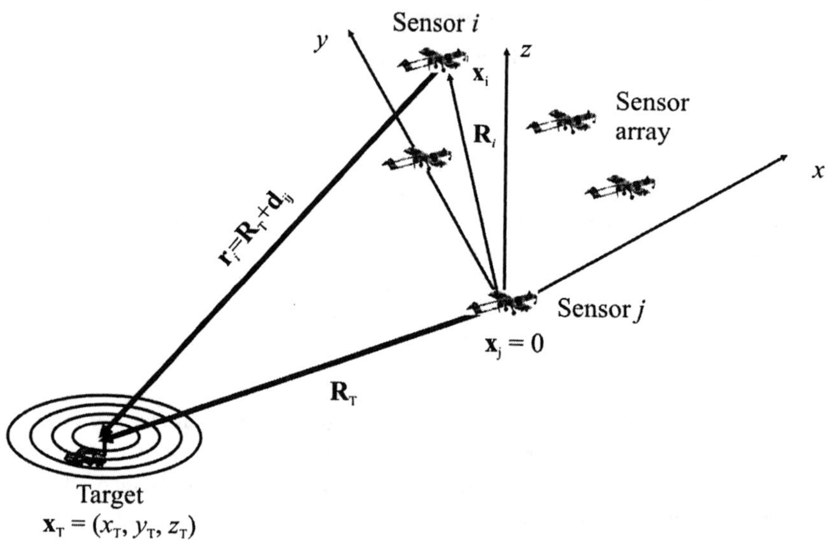

Figure 3.32 Geometry of the problem. (*Source:* [24], © 1987 IEEE. Reprinted with permission.)

$$x_j = 0 \quad \text{and} \quad R_j = 0, r_j = R_\mathrm{T} \tag{3.141}$$

Now

$$(R_\mathrm{T} + r_{ij})^2 = R_i^2 - 2\mathbf{x}_i^\mathrm{T}\mathbf{x}_\mathrm{T} + R_\mathrm{T}^2 \tag{3.142}$$

so

$$R_i^2 - r_{ij}^2 - 2R_\mathrm{T}r_{ij} - 2\mathbf{x}_i^\mathrm{T}\mathbf{x}_T = 0 \tag{3.143}$$

yielding $N-1$ equations in three unknowns.

With measurement errors and noise in the data, (3.143) is not identically zero, but it is equal to some residual amount of error denoted by ε_i. That is,

$$R_i^2 - r_{ij}^2 - 2R_\mathrm{T}r_{ij} - 2\mathbf{x}_i^\mathrm{T}\mathbf{x}_\mathrm{T} = \varepsilon_i, \qquad i = 1, 2, \ldots, N \tag{3.144}$$

In matrix notation, (3.144) can be expressed as

$$\boldsymbol{\varepsilon} = \boldsymbol{\delta} - 2R_\mathrm{T}\mathbf{r} - 2\mathbf{S}\mathbf{x}_\mathrm{T} \tag{3.145}$$

where

$$\boldsymbol{\delta} = \begin{bmatrix} R_2^2 - r_{21}^2 \\ R_3^2 - r_{31}^2 \\ \vdots \\ R_N^2 - r_{N1}^2 \end{bmatrix} \qquad \mathbf{r} = \begin{bmatrix} r_{21} \\ r_{31} \\ \vdots \\ r_{N1} \end{bmatrix} \qquad \mathbf{S} = \begin{bmatrix} x_2 & y_2 & z_2 \\ x_3 & y_3 & z_3 \\ \vdots & \vdots & \vdots \\ x_N & y_N & z_N \end{bmatrix}$$

The least-squared solution for $\mathbf{x}_\mathrm{T}$ when R_T is known with weighted equation error, where the weights $\mathbf{W}$ on the range differences are determined a priori, is

$$\mathbf{x}_T = \frac{1}{2}\mathbf{S}_W^*(\boldsymbol{\delta} - 2R_T\mathbf{r}) \tag{3.146}$$

where $\mathbf{S}_W^*$ is given by (2.16),

$$\mathbf{S}_W^* = (\mathbf{S}^\mathrm{T}\mathbf{W}\mathbf{S})^{-1}\mathbf{S}^\mathrm{T}\mathbf{W} \tag{3.147}$$

Unfortunately, R_T is unknown if the location of the target is unknown, which it is in this case. In addition, expression (3.146) is nonlinear and some nonlinear minimization technique must be used to solve it.

Smith and Abel proposed a technique they referred to as the *spherical interpolation* (SI) method for solving (3.146). Furthermore they compare three types of PF techniques based on range difference computations:

- The SI method;
- The *spherical intersection* (SX) method;
- The *plane intersection* (PX) method.

The SX method was developed by Schau and Robinson [27] and the PX method is based on a development by Schmidt described later.

3.3.2.1 Spherical Interpolation Method

The SI method is based on substituting (3.146) into (3.145) and minimizing $\boldsymbol{\varepsilon}$ with respect to R_T. First, two matrices are defined as

$$P_{s,N-1\times N-1} \triangleq \mathbf{S}(\mathbf{S}^\text{T}\mathbf{W}\mathbf{S})^{-1}\mathbf{S}^\text{T}\mathbf{W} \tag{3.148}$$

and

$$\mathbf{P}^{\perp}_{s,N-1\times N-1} = \mathbf{I} - \mathbf{P}_s \tag{3.149}$$

yielding

$$\tilde{R}_\text{T} = \frac{1}{2}\frac{\mathbf{r}^\text{T}\mathbf{P}_s^{\perp}\mathbf{V}\mathbf{P}_s^{\perp}\boldsymbol{\delta}}{\mathbf{r}^\text{T}\mathbf{P}_s^{\perp}\mathbf{V}\mathbf{P}_s^{\perp}\mathbf{r}} \tag{3.150}$$

where $\mathbf{V}$ is a symmetric positive definite weighting matrix.

The resulting estimate for $\mathbf{x}_\text{T}$ that minimizes the error in an LS sense is given by

$$\hat{\mathbf{x}}_\text{s} = \frac{1}{2}(\mathbf{S}^\text{T}\mathbf{P}_d^{\perp}\mathbf{W}\mathbf{P}_d^{\perp}\mathbf{S})^{-1}\mathbf{S}^\text{T}\mathbf{P}_d^{\perp}\mathbf{W}\mathbf{P}_d^{\perp}\boldsymbol{\delta} \tag{3.151}$$

where

$$\mathbf{P}_d^{\perp} \triangleq \mathbf{I} - \frac{\mathbf{r}\mathbf{r}^{\mathrm{T}}}{\mathbf{r}^{\mathrm{T}}\mathbf{r}} \tag{3.152}$$

is a projection matrix.

3.3.2.2 Spherical Intersection Method

The spherical intersection solution to find $\mathbf{x}_{\mathrm{T}}$ is found by substituting (3.146) into $R_{\mathrm{T}}^2 = \mathbf{x}_{\mathrm{T}}^{\mathrm{T}}\mathbf{x}_{\mathrm{T}}$, which yields

$$aR_{\mathrm{T}}^2 + bR_{\mathrm{T}} + c = 0 \tag{3.153}$$

where

$$a = 4 - 4\mathbf{r}^{\mathrm{T}}\mathbf{S}_W^{*\,\mathrm{T}}\mathbf{S}_W^{*}\mathbf{r} \tag{3.154}$$

$$b = 4\mathbf{r}^{\mathrm{T}}\mathbf{S}_W^{*\,\mathrm{T}}\mathbf{S}_W^{*}\boldsymbol{\delta} \tag{3.155}$$

$$c = -\boldsymbol{\delta}^{\mathrm{T}}\mathbf{S}_W^{*\,\mathrm{T}}\mathbf{S}_W^{*}\boldsymbol{\delta} \tag{3.156}$$

Then (3.153) yields two solutions as given by the roots

$$R_{\mathrm{T}} = \frac{-b \pm \sqrt{b^2 - 4ac}}{2a} \tag{3.157}$$

Given R_{T}, it is substituted into (3.146) to obtain $\mathbf{x}_{\mathrm{T}}$.

3.3.2.3 Plane Intersection Method

This technique is based on a development due to Schmidt [28]. In this case,

$$\varepsilon_{i\,jk} = 2\boldsymbol{\Delta}_{ijk}^{\mathrm{T}}\mathbf{S}_{ijk}\mathbf{x}_{\mathrm{T}} - [r_{ji}r_{kj}r_{ik} + R_i^2 r_{kj} + R_j^2 r_{ik} + R_k^2 r_{ji}] \tag{3.158}$$

where $\varepsilon_{i\,jk}$ is the equation error, r_{jk} is the range difference measured between sensors j and k, and

$$R_i^2 = x_i^2 + y_i^2 \tag{3.159}$$

where $\mathbf{x}_i = [x_i \quad y_i]^{\mathrm{T}}$ when x_i and y_i are the coordinates of sensor S_i. The three coordinate terms in (3.158) are given by

$$\boldsymbol{\Delta}_{ijk} = \begin{bmatrix} r_{kj} & r_{ik} & r_{ji} \end{bmatrix}^{\mathrm{T}} \tag{3.160}$$

$$\mathbf{S}_{ijk} \triangleq \begin{bmatrix} \mathbf{x}_i^{\mathrm{T}} & \mathbf{x}_j^{\mathrm{T}} & \mathbf{x}_k^{\mathrm{T}} \end{bmatrix}^{\mathrm{T}} \tag{3.161}$$

Expression (3.158) is linear in $\mathbf{x}_{\mathrm{T}}$ and therefore can be used as the basis for an LS solution, just as (3.144) was used above.

3.3.2.4 Performance Comparison

To compare the performance of these three methods, Smith and Abel conducted Monte-Carlo simulation experiments with the configuration shown in Figure 3.33 and Table 3.1. One hundred runs with four configurations were performed. The four cases were according to the parameters shown in Table 3.1. In all cases the sensors were located at (0, 0, 0), (0, 0, 100), (0, 0, 200), (100, 0, 0), (100, 0, 100), (100, 0, 200), (0, 100, 0), (0, 100, 100), and (0, 100, 200).

The bias results are given in Table 3.2, while the sample standard deviations are given in Table 3.3. The sample *root-mean-squared* (RMS) values are given in Table 3.4. The spherical intersection is seen to have much higher errors than the

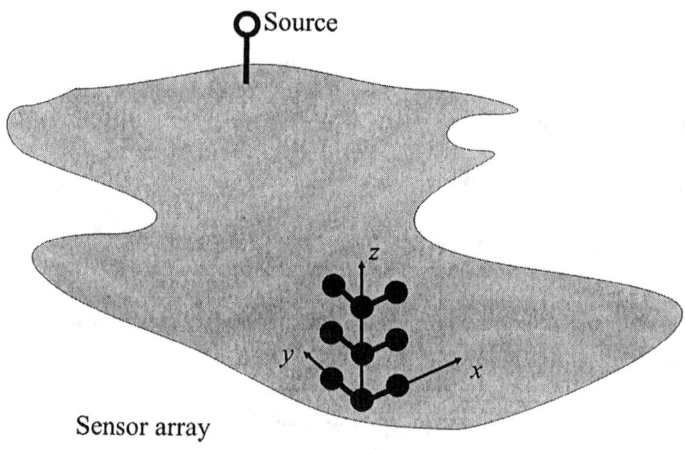

Figure 3.33 Range difference example. (*Source:* [24], © 1987 IEEE. Reprinted with permission.)

Table 3.1 Simulation Cases

Case	Target location $\mathbf{x}_T$	Range R_T	Bearing cosines Ω_s	RD noise standard deviation σ_{di1}
1	(390, 160, 170)	454.5	(0.858, 0.352, 0.374)	0.1
2	(390, 160, 170)	454.5	(0.858, 0.352, 0.374)	1.0
3	(540, 1,360, 110)	1,467	(0.368, 0.927, 0.750)	0.1
4	(540, 1,360, 110)	1,467	(0.368, 0.927, 0.750)	1.0

Source: [24].

Table 3.2 Biases in the Computed PFs

Method	Run	Source location bias $\hat{x}_T - x_T$	$\hat{y}_T - y_T$	$\hat{z}_T - z_T$	Range $\hat{R}_T - R_T$	Bearing $\hat{\theta}_T - \theta_T$
Spherical Interpolation	1	0.239	0.60	0.026	0.225	0.010
	2	5.16	1.47	0.837	5.42	0.174
	3	1.87	4.60	0.003	4.95	0.014
	4	62.7	165	1.40	178	0.500
Plane Intersection	1	0.195	0.052	−0.035	0.183	0.010
	2	4.24	1.22	0.602	4.19	0.150
	3	0.961	2.18	−0.035	2.37	0.009
	4	26.9	72.5	1.8	76.9	0.2
Spherical Intersection	1	0.341	0.090	0.048	0.341	0.012
	2	0.316	−0.129	−0.057	0.129	0.035
	3	−2.0	−5.76	−0.090	−6.10	0.017
	4	*	*	*	*	*

Source: [24].

* Results too large to be useful.

Table 3.3 Standard Deviations of the Simulations

Method	Run	Source location error $\sigma_{\hat{x}_T}$	$\sigma_{\hat{y}_T}$	$\sigma_{\hat{z}_T}$	Range $\sigma_{\hat{R}_T}$	Bearing $\sigma_{\hat{\theta}_T}$
Spherical Interpolation	1	2.07	0.698	0.445	2.27	0.042
	2	20.4	6.91	4.39	22.4	0.41
	3	9.98	26.2	0.600	28.2	0.07
	4	82	219	5	234	0.7
Plane Intersection	1	4.17	1.33	0.898	4.37	0.07
	2	40.3	12.8	8.732	42.2	0.66
	3	15.8	42.5	0.817	45.2	0.11
	4	159	421	7	449	1.0
Spherical Intersection	1	3.12	1.10	0.513	3.24	0.066
	2	31.1	11.1	5.13	32.4	0.64
	3	39.4	105	0.958	112	0.28
	4	*	*	*	*	*

Source: [24].
*Results too large to be useful.

Table 3.4 RMS Values Simulation Results

Method	Run	Source location error $\bar{\sigma}_{\hat{x}_T}$	$\bar{\sigma}_{\hat{y}_T}$	$\bar{\sigma}_{\hat{z}_T}$	Range $\bar{\sigma}_{\hat{R}_T}$	Bearing $\bar{\sigma}_{\hat{\theta}_T}$
Spherical Interpolation	1	2.08	0.700	0.446	2.28	0.043
	2	21	7.06	4.47	23	0.45
	3	10.1	26.6	0.600	28.6	0.071
	4	103	274	5.2	294	0.7
Plane Intersection	1	4.18	1.33	0.898	4.37	0.07
	2	40.5	12.9	8.75	42.4	0.68
	3	15.8	42.6	0.818	45.3	0.11
	4	161	427	7	456	1
Spherical Intersection	1	3.12	1.1	0.515	3.25	0.067
	2	31.1	11.1	5.13	32.4	0.64
	3	39.4	105	0.962	112	0.28
	4	*	*	*	*	*

Source: [24].
*Results too large to be useful.

other two methods. The spherical interpolation has the lowest RMS error, followed not too distantly by the plane intersection method.

3.3.2.5 Geometrical Interpretations

The geometrical arguments put forth by Smith and Abel are particularly interesting and point out some physical reasons why one technique performs better than another.

The SI geometry is shown in Figure 3.34. The RDs in this case are spread over a reasonable range. This will make the technique relatively insensitive to noise-induced errors.

The SX geometry is shown in Figure 3.35. The target is located at the intersection of spheres with radius $\hat{R}_T + r_{i1}$ around each sensor. When the target is far away from the sensor array, these spheres have large radii and their intersection is close to the intersection of parallel lines—the intersection could be anywhere on a long circle segment. This leads to inherent inaccuracies as the SNR is decreased.

For the PX method, the target lies on the major axis of some conical shape. Shown in Figure 3.36 are two ellipses. Considering three sensors at a time, the three sensors define a plane. The conic shape is on this plane and therefore so is its major axis.

With two such ellipses, the major axes intersect at a point and that, of necessity, is where the target lies. With more than two ellipses and including measurement error and noise, these conic axes do not intersect at a point and therefore the LS process is required. Nevertheless, the intersection of the axes indicating the location of the target produces reasonable crossing angles, so higher noise immunity would be expected.

3.3.3 Range Difference Using Feasible Bivectors

A novel approach to geolocation with range differences was developed by Schmidt [28]. The development included use of Grassmann algebra, some details of which are given in Appendix A.

The approach is based on modifying each set of TDOAs from three sensors so that their sum is zero by subtracting their average. Each of these sets defines a plane which may or may not (mostly will not) contain the location of the target. The planes will not normally contain the target due to noise in the process and the intersection of the planes (the intersection of two planes defines a straight line and the intersection of three planes defines a point). The target will, however, be close to these planes. Once these TDOAs are thus modified, a least-squared computation

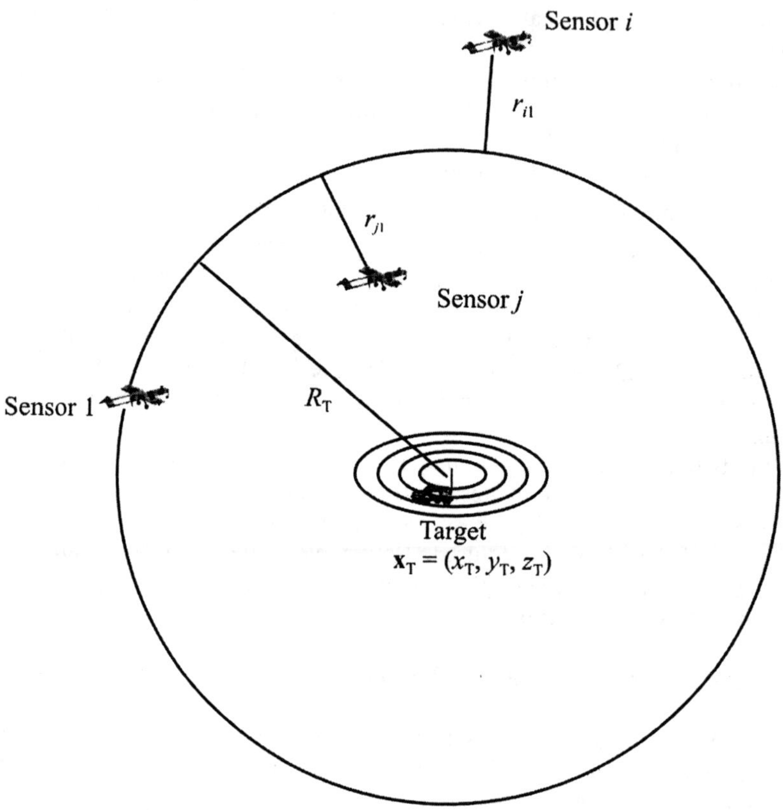

Figure 3.34 The SI configuration of the sensors and target geometry. (*From:* [24], © 1987 IEEE. Reprinted with permission.)

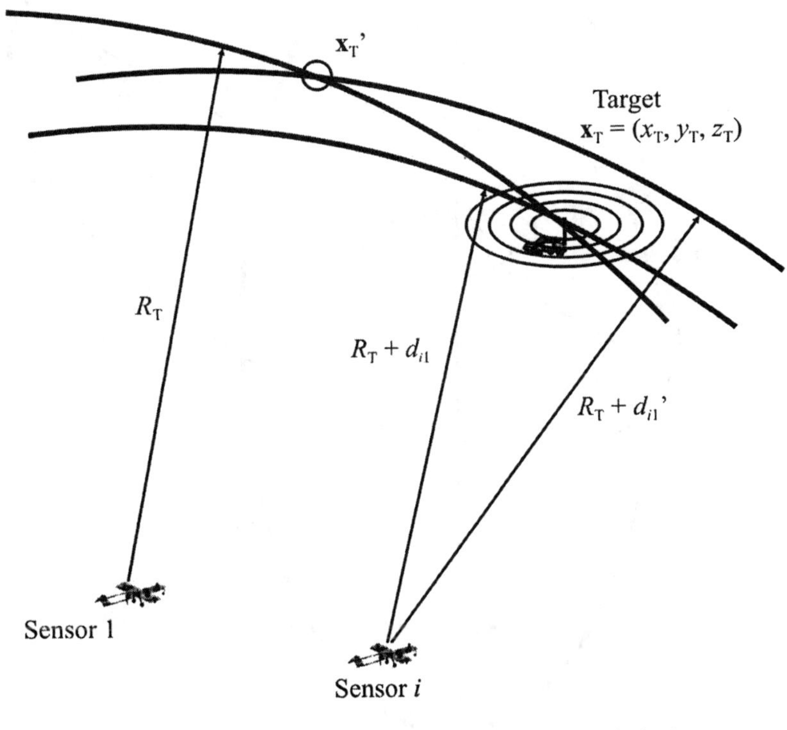

Figure 3.35 SX geometry. (*Source:* [24], © 1987 IEEE. Reprinted with permission.)

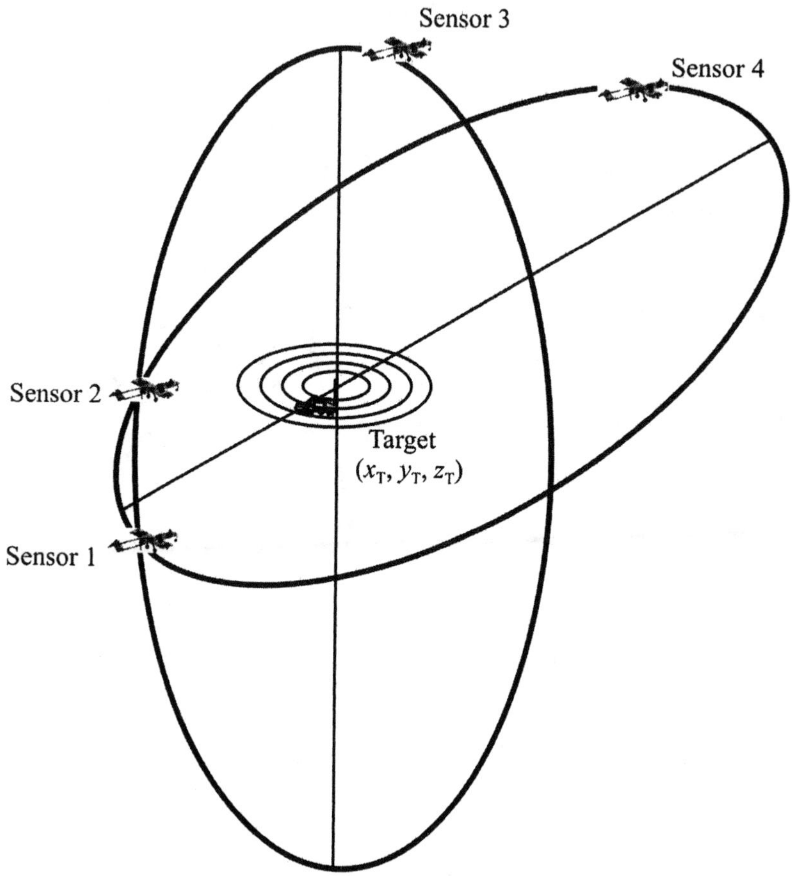

Figure 3.36 PX geometry.

finds the point that is closest to all the planes. That is the estimated location of the target.

Given that there are N sensors with the coordinates of the ith sensor being (x_i, y_i, z_i), then

$$r_i = \sqrt{(x_\mathrm{T} - x_i)^2 + (y_\mathrm{T} - y_i)^2 + (z_\mathrm{T} - z_i)^2} \tag{3.162}$$

is the range between the ith sensor and the target when $(x_\mathrm{T}, y_\mathrm{T}, z_\mathrm{T})$ are the coordinates of the target emitter. The range differences are given by

$$\Delta_{ij} = r_j - r_i, \qquad 1 \le i, j \le N \tag{3.163}$$

and $\Delta_{ji} = -\Delta_{ij}$, as well as

$$\Delta_{ii} = 0$$

A differencing matrix **D** is defined such that

$$\mathbf{\Delta} = \mathbf{Dr} \tag{3.164}$$

For N sensors it is denoted $\mathbf{D}_N$. Thus, for $N = 4$,

$$\begin{bmatrix} \Delta_{12} \\ \Delta_{13} \\ \Delta_{14} \\ \Delta_{23} \\ \Delta_{24} \\ \Delta_{34} \end{bmatrix} = \begin{bmatrix} -1 & 1 & 0 & 0 \\ -1 & 0 & 1 & 0 \\ -1 & 0 & 0 & 1 \\ 0 & -1 & 1 & 0 \\ 0 & -1 & 0 & 1 \\ 0 & 0 & -1 & 1 \end{bmatrix} \begin{bmatrix} r_1 \\ r_2 \\ r_3 \\ r_4 \end{bmatrix} \tag{3.165}$$

Vector $\mathbf{\Delta}$ is referred to as a *bivector*, since each element has two subscripts, not just one as in a linear vector. As noted in [29], $\mathbf{\Delta}$ cannot be just any bivector; it must lie in the range space of **D**. If $\mathbf{\Delta}$ can be expressed in the form of (3.164) it is said to be *feasible*, otherwise it is *infeasible*.

Given a real (measured) bivector $\mathbf{\Delta}$, the nearest feasible bivector $\hat{\mathbf{\Delta}}$ is obtained in an LS sense by projecting $\mathbf{\Delta}$ onto the range space of **D**. That is,

$$\hat{\mathbf{\Delta}} = \mathbf{P\Delta} \tag{3.166}$$

The projection vector is

$$\mathbf{P} = \mathbf{D}(\mathbf{D}^{\mathrm{T}}\mathbf{D})^{-}\mathbf{D}^{\mathrm{T}} \tag{3.167}$$

where $\mathbf{D}^{-}$ is the generalized inverse of $\mathbf{D}$[1]. In addition, in this case,

$$\mathbf{D}^{\mathrm{T}}\mathbf{D} = N\left(\mathbf{I} - \frac{1}{N}\mathbf{1}\mathbf{1}^{\mathrm{T}}\right) \tag{3.168}$$

where $\mathbf{1}$ is a vector of all 1's. The term in the brackets in (3.168) is its own generalized inverse, so

$$\begin{aligned}
\mathbf{P} &= \mathbf{D}(\mathbf{D}^{\mathrm{T}}\mathbf{D}) - \mathbf{D}^{\mathrm{T}} = \mathbf{D}(N\mathbf{I} - \mathbf{1}\mathbf{1}^{\mathrm{T}}) - \mathbf{D}^{\mathrm{T}} \\
&= \frac{1}{N}\mathbf{D}\left(\mathbf{I} - \frac{1}{N}\mathbf{1}\mathbf{1}^{\mathrm{T}}\right)\mathbf{D}^{\mathrm{T}} \\
&= \frac{1}{N}\mathbf{D}\mathbf{D}^{\mathrm{T}} - \frac{1}{N^2}\mathbf{D}\mathbf{1}\mathbf{1}^{\mathrm{T}}\mathbf{D}^{\mathrm{T}} && (3.169) \\
&= \frac{1}{N}\mathbf{D}\mathbf{D}^{\mathrm{T}} && (3.170)
\end{aligned}$$

because the second term in (3.169) is zero.

Any *algebra* is a mathematical construct that includes vectors and an operation called the product that is closed under that operation. In Grassmann algebra, the product operation is called the *wedge product* and is normally denoted by $\wedge$. "Closed" means that if $\mathbf{a}$ and $\mathbf{b}$ are vectors in the algebra, then so is $\mathbf{a} \wedge \mathbf{b}$. Note that the set of bivectors generated by a set of vectors is not in the same space as the vectors and therefore does not form an algebra. Extending the results, however, to include all higher forms does form an algebra. If $N = 4$, then the wedge product of $\mathbf{x} = [x_1 \ x_2 \ x_3 \ x_4]^{\mathrm{T}}$ and $\mathbf{y} = [y_1 \ y_2 \ y_3 \ y_4]^{\mathrm{T}}$ is given by

[1] A generalized inverse of matrix $\mathbf{A}$ is any matrix $\mathbf{B}$ satisfying $\mathbf{ABA} = \mathbf{A}$.

$$x \wedge y = \begin{vmatrix} x_1 & y_1 \\ x_2 & y_2 \\ x_3 & y_3 \\ x_4 & y_4 \end{vmatrix} = \begin{bmatrix} x_1 y_2 - x_2 y_1 \\ x_1 y_3 - x_3 y_1 \\ x_1 y_4 - x_4 y_1 \\ x_2 y_3 - x_3 y_2 \\ x_2 y_4 - x_4 y_2 \\ x_3 y_4 - x_4 y_2 \end{bmatrix} = \begin{bmatrix} z_{12} \\ z_{13} \\ z_{14} \\ z_{23} \\ z_{24} \\ z_{34} \end{bmatrix} \tag{3.171}$$

Note that

$$\mathbf{1} \wedge \mathbf{r} = \begin{vmatrix} 1 & r_1 \\ 1 & r_2 \\ 1 & r_3 \\ 1 & r_4 \end{vmatrix} = \begin{bmatrix} r_2 - r_1 \\ r_3 - r_1 \\ r_4 - r_1 \\ r_3 - r_2 \\ r_4 - r_2 \\ r_4 - r_3 \end{bmatrix} = \begin{bmatrix} \Delta_{21} \\ \Delta_{31} \\ \Delta_{41} \\ \Delta_{32} \\ \Delta_{42} \\ \Delta_{43} \end{bmatrix} = \boldsymbol{\Delta} \tag{3.172}$$

Adding a second product to (3.172) forces the product to zero as

$$\mathbf{1} \wedge \boldsymbol{\Delta} = \mathbf{1} \wedge \mathbf{1} \wedge \mathbf{r} = \mathbf{0} \tag{3.173}$$

This expression is true because any bivector multiplied by itself is zero; to wit,

$$\mathbf{1}_3 \wedge \mathbf{1}_3 = \begin{vmatrix} 1 & 1 \\ 1 & 1 \\ 1 & 1 \end{vmatrix} = \begin{vmatrix} 1-1 \\ 1-1 \\ 1-1 \end{vmatrix} = \mathbf{0}_3 \tag{3.174}$$

where $\mathbf{1}_N$ is a vector of all 1's of length N. Expression (3.173) is called a *trivector equation*. Expression (3.173) is true if and only if $\boldsymbol{\Delta}$ is a feasible bivector. With measurement errors and noise included, (3.173) will not be identically true and the closest feasible bivector $\hat{\boldsymbol{\Delta}}$ is sought.

The trivector equation $\mathbf{1} \wedge \boldsymbol{\Delta} = \mathbf{s}$, including measurement errors and noise, can be expressed in terms of vectors and matrices as $\mathbf{S}\boldsymbol{\Delta} = \mathbf{s}$. For $N = 4$, for example,

$$\mathbf{S}_4\,\mathbf{\Delta} = \begin{bmatrix} 1 & -1 & 0 & 1 & 0 & 0 \\ 1 & 0 & -1 & 0 & 1 & 0 \\ 0 & 1 & -1 & 0 & 0 & 1 \\ 0 & 0 & 0 & 1 & -1 & 1 \end{bmatrix} \begin{bmatrix} \Delta_{12} \\ \Delta_{13} \\ \Delta_{14} \\ \Delta_{23} \\ \Delta_{24} \\ \Delta_{34} \end{bmatrix} = \begin{bmatrix} s_{123} \\ s_{124} \\ s_{134} \\ s_{234} \end{bmatrix} \tag{3.175}$$

The matrix product $\mathbf{S}_n\mathbf{\Delta}$ equals zero if $\mathbf{\Delta}$ is feasible and equals the trivector of residuals if not.

The matrices $\mathbf{D}_n$ and $\mathbf{S}_n$ can be found recursively by

$$\mathbf{D}_{N+1} = \begin{bmatrix} -\mathbf{1}_N & \mathbf{1}_N \\ \mathbf{0} & \mathbf{D}_N \end{bmatrix} \qquad \mathbf{S}_{N+1} = \begin{bmatrix} -\mathbf{D}_N & \mathbf{I}_{\binom{N}{2}} \\ \mathbf{0} & \mathbf{S}_N \end{bmatrix} \tag{3.176}$$

Schmidt defines TDOA (range difference) averaging as

$$\hat{\Delta}_{ij} = \Delta_{ij} - \frac{1}{n} \sum_{k \neq i,j} (\Delta_{ij} + \Delta_{jk} + \Delta_{ki}) \qquad 1 \le i, j \le n \tag{3.177}$$

which simply subtracts from each triad measured range difference the average of the range differences involved. The term in brackets is referred to as *circuital sum* and will equal zero if Δ_{ij} is feasible (lies on a plane).

It is also proven in [28] that TDOA (range difference) averaging produces the same result as minimum least-squared estimating. The algorithm for emitter geolocation is given as the following:

1. Given the measured range difference bivector $\mathbf{\Delta}$, calculate the closest feasible range difference bivector $\hat{\mathbf{\Delta}}$ using either (3.177) or (3.166), since they are equivalent.
2. Calculate the point nearest to the $\binom{N}{3}$ planes of position, one for each of the triads of elements of $\hat{\mathbf{\Delta}}$.

The second step can be executed using the *location on the conic axis* (LOCA) algorithm developed in [29]. It is the algorithm upon which (3.158) is based and is summarized by the following:

> Given that sensor i is located at (x_i, y_i, z_i) and there are N sensors total, then the target lies in planes described by
>
> $$Ax + By + Cz = D \tag{3.178}$$
>
> where
>
> $$\begin{aligned} A &= x_1\Delta_{23} + x_2\Delta_{31} + x_3\Delta_{12} \\ B &= y_1\Delta_{23} + y_2\Delta_{31} + y_3\Delta_{12} \\ C &= z_1\Delta_{23} + z_2\Delta_{31} + z_3\Delta_{12} \\ D &= \frac{1}{2}(\Delta_{12}\Delta_{23}\Delta_{31} + d_1^2\Delta_{23} + d_2^2\Delta_{31} + d_3^2\Delta_{12}) \\ d_i &= \sqrt{x_i^2 + y_i^2 + z_i^2} \end{aligned}$$
>
> For N sensors there are $\binom{N}{3}$ such planes. Thus,
>
> $$\begin{bmatrix} A_{123} & B_{123} & C_{123} \\ \vdots & \vdots & \vdots \\ A_{ijk} & B_{ijk} & C_{ijk} \\ \vdots & \vdots & \vdots \end{bmatrix} \begin{bmatrix} x \\ y \\ z \end{bmatrix} = \begin{bmatrix} D_{123} \\ \vdots \\ D_{ijk} \\ \vdots \end{bmatrix} \tag{3.179}$$
>
> The point closest to the intersection of these planes provides the least-squared estimate of the target location.

Schmidt presents simulation results that are based on the same geometry as that used by Smith and Abel. Those results are given in Tables 3.5 and 3.6, but the SX results have been omitted, since those results are much worse than the rest.

Comparing the results in Table 3.6 with those in Tables 3.2 through 3.4 indicates that the feasible bivector method is the best of those analyzed. The improvements range from a factor of just a few to over 10 times better.

Table 3.5 Least-squared and Weighted Least-squared Simulation Results

Run	ε_x	σ_x	ε_y	σ_y	ε_z	σ_z	RMS
			LS Plane Intersection				
1	−0.03	1.6	0.0005	0.63	−0.002	0.32	1.2
2	−7.7	15.6	−2.8	5.8	−1.2	3.5	18.3
3	−6.2	7.4	−16.7	20	−0.06	0.66	17.5
4	−226	104	−610	283	−4.1	3.0	690
			Weighted LS Plane Intersection				
1	0.06	1.1	0.01	0.37	0.0002	0.23	1.30
2	−8.3	11.7	−2.7	4.3	−1.6	2.6	13.2
3	−1.2	5.1	−3.4	13.8	−0.05	0.30	14.5
4	−102	27	−276	74.5	−1.6	2.4	156

Source: [28].

3.4 Concluding Remarks

Calculating target geolocation estimates by methods other than triangulation were presented in this chapter. These techniques are based on calculating the TDOA, DD, and range differences between several sensors. These methods invariably lead to quadratic equations to solve for the PF.

The principal advantage of these techniques over triangulation is the antenna configuration—quadratic methods typically require only a single antenna, while triangulation typically requires an antenna array. The biggest drawback is that for many communication signals there is no feature that can be used to correlate arrival times between sensors. This leads to the requirement to compute the pre-detect cross-correlation function of the received signals, which requires exchanging large amounts of data leading to the requirement for significant data link capacity.

The quadratic methods generally produce more accurate PFs than triangulation. Among the quadratic methods discussed here, the RD method using feasible bivector calculations based on Grassmann algebra is the most accurate.

The quadratic methods are still subject to GDOP-induced errors, however.

Table 3.6 Feasible Bivector Simulation Results

Run	ε_x	σ_x	ε_y	σ_y	ε_z	σ_z	RMS
1	−0.02	0.53	−0.01	0.19	0.004	0.12	0.678
2	0.15	5.6	0.10	2.0	0.08	1.3	6.79
3	0.55	3.7	1.7	10.0	−0.01	0.20	0.37
4	−9.4	35	−25	95	−0.13	2.2	93.75

Source: [28].

They are also subject to the errors produced by noise and inaccurate measurements.

References

[1] Poisel, R. A., *Introduction to Communication Electronic Systems,* Norwood, MA: Artech House, 2002, Chapter 12.

[2] Rusu, P., "The Equivalence of TOA and TDOA RF Transmitter Location," Applied Research Laboratories, the University of Texas at Austin.

[3] Shin, D-H., and T-K. Sung, "Comparison of Error Characteristics between TOA and TDOA Positioning," *IEEE Transactions on Aerospace and Electronic Systems,* Vol. 38, No. 1, January 2002, pp. 307–310.

[4] Bard, J. D., F. M. Ham, and W. L. Jones, "An Algebraic Solution to the Time Difference of Arrival Equations," *Proceedings of the IEEE SouthEastcon Conference,* Tampa, April 11–14, 1996, pp. 313–319.

[5] Mellen G. M. II, Pachter, and J. Raquet, "Closed-Form Solution for Determining Emitter Location Using Time Difference of Arrival Measurements," *IEEE Transactions on Aerospace and Electronic Systems,* Vol. 39, No. 3, July 2003, pp. 1056–1058.

[6] Piersol, A. G., "Time Delay Estimation Using Phase Data," *IEEE Transactions on Acoustics, Speech, and Signal Processing,* Vol. ASSP-29, No. 3, June 1981, pp. 471–477.

[7] Stein, S., "Algorithms for Ambiguity Functions Processing," *IEEE Transactions on Acoustic, Speech, and Signal Processing,* Vol. ASSP-29, No. 3, June 1981, pp. 588–599.

[8] Quazi, A. H., "An Overview on the Time Delay Estimate in Active and Passive Systems for Target Localization," *IEEE Transactions on Acoustics, Speech, and Signal Processing,* Vol. ASSP-29, No. 3, June 1981, pp. 527–533.

[9] Knapp, C. H., and G. C. Carter, "The Generalized Correlation Method for Estimation of Time Delay," *IEEE Transactions on Acoustics, Speech, and Signal Processing,* Vol. ASSP-24, August 1976, pp. 320–327.

[10] Schultheiss, P. M., "Locating a Passive Source with Array Measurements—A Summary Result," *Proceedings ICASSP 1979,* Washington, D. C., April 24, 1979, pp. 967–970.

[11] Hahn, W. R., "Optimum Signal Processing for Passive Sonar Range and Bearing Estimation," *Journal Acoustic Society of America,* Vol. 58, July 1975, pp. 201–207.

[12] Tomlinson, G., and J. Sorokowsky, "Accuracy Prediction Report, Raploc Error Model," IBM, Manassas, VA, June 30, 1977, pp. B1–B7.

[13] Gustafsson, F., and F. Gunnarsson, "Positioning Using Time-Difference of Arrival Measurements," *Proceedings IEEE International Conference on Acoustics, Speech, and Signal Processing, ICAASP 03,* Vol. 6, Hong Kong, April 6–10 2003, pp. VI-553–556.

[14] Nordlund, P. J., F. Gunnarsson, and F. Gustafsson, "Particle Filters for Positioning in Wireless Networks," *EUSIPCO,* Toulouse, September 2002.

[15] Arulampalam, M. S. et al., "A Tutorial on Particle Filters for Online Nonlinear/Non-Gaussian Baysian Tracking," *IEEE Transactions on Signal Processing,* Vol. 50, No. 2, February 2002, pp. 174–188.

[16] Gustafsson, F., et al., "Particle Filters for Positioning, Navigation, and Tracking," *IEEE Transactions on Signal Processing,* Vol. 50, No. 2, February 2002, pp. 425–437.

[17] Bard, J. D., and F. M. Ham, "Time Difference of Arrival Dilution of Precision and Applications," *IEEE Transactions on Signal Processing,* Vol. 47, No. 2, February 1999, pp. 521–523.

[18] Hofman-Wellenhof, B., H. Lichtenegger, and J. Collins, *GPS Theory and Practice,* 3rd Ed., Wein, NY: Springer-Verlag, 1994, pp. 235–237, 249–253.

[19] Koorapaty, H., H. Grubeck, and M. Cedervall, "Effect of Biased Measurement Errors on Accuracy of Position Location Methods," *IEEE GLOBECOM 98,* Vol. 3, 8–12 Nov. 1998, pp. 1497–1502.

[20] Chan, Y. T., and K. C. Ho, "TDOA-SDOA Estimation with Moving Source and Receivers," *IEEE Transactions ICASSP 2003,* 2003, pp. V-153–V-156.

[21] Levanon, N., "Interferometry against Differential Doppler: Performance Comparison of Two Emitter Location Airborne Systems," *IEE Proceedings,* Vol. 136, Pt. F., No. 2, April 1989, pp. 70–74.

[22] Otnes, R. K., "Frequency Difference of Arrival Accuracy," *IEEE Transactions on Acoustic, Speech, and Signal Processing,* Vol. 37, No. 2, February 1989, pp. 306–308.

[23] Ullman, R. J., and E. Geraniotis, "Motion Detection Using TDOA and FDOA Measurements," *IEEE Transactions on Aerospace and Electronic Systems,* Vol. 37, No. 2, April 2001, pp. 759–764.

[24] Rusu, P. I., "A TOA-FOA Solution for the Position and Velocity of RF Emitters," Applied Research Laboratories Report #SGL-GR-99-15, The University of Texas at Austin.

[25] Ho, K.C., and W. Xu, "An Accurate Algebraic Solution for Moving Source Location Using TDOA and FDOA Measurements," *IEEE Transactions on Signal Processing,* Vol. 52, No. 9, September 2004, pp. 2453–2463.

[26] Smith, J. O., and J. S. Abel, "Closed-Form Least-Squares Source Location Estimation from Range-Difference Measurements," *IEEE Transactions on Acoustics, Speech, and Signal Processing*, Vol. ASSP-35, No. 2, December 1987, pp. 1661–1669.

[27] Schau, H. C., and A. Z. Robinson, "Passive Source Localization Employing Intersecting Spherical Surfaces from Time-of-Arrival Differences," *IEEE Transactions on Acoustics, Speech, and Signal Processing*, Vol. ASSP-35, No. 8, August 1987, pp. 1223–1225.

[28] Schmidt, R. O., "Least-Squares Range Difference Location," *IEEE Transactions on Aerospace and Electronic Systems,* Vol. 32, No. 1, January 1996, pp. 234–242.

[29] Schmidt, R. O., "A New Approach to Geometry of Range Difference Location," *IEEE Transactions on Aerospace and Electronic Systems,* Vol. AES-8, November 1972, pp. 821–835.

Chapter 4

Single-Site Location Techniques

In this chapter we will discuss techniques for geolocating HF signals using only a single-sensor site. As will be seen, this is possible because long-range HF signals are reflected (refracted actually) in the ionosphere back to the Earth, and measuring the angles of arrival in three dimensions allows calculation of the PF of the transmitter when the height of the point of reflection is known. This height information is not required, however, when there are two or more ray paths arriving at the sensor sight and the angles of arrival of these rays can be measured and discerned.

4.1 HF Signal Propagation

Under some circumstances, HF signals can propagate for considerable distances. This long-range communication capability of such signals was used by navies around the world for ship-to-shore and ship-to-ship communications before satellite systems were available. In fact, it is still used for that purpose.

The ionospheric layers that surround the Earth are responsible for the phenomenon. During the daytime there are four layers involved in the process, the lowest being the D-layer, the next the E-layer, and then two F-layers: the F1 and F2. At night the two F-layers combine into one and the D- and E-layers disappear, as shown in Figure 4.1. These layers are regions of ions that have been charged by bombardment of atoms with photons from the sun and the resulting free electrons. It is widely believed that it is the free electrons that are responsible for the refraction phenomenon; the positively charged ions have a minor role. They interact with HF signals and the Earth's magnetic field in such a way that *electromagnetic* (EM) wave refraction is possible. Normally this is below 10 MHz or so, but can occur higher depending on the density of the free electrons.

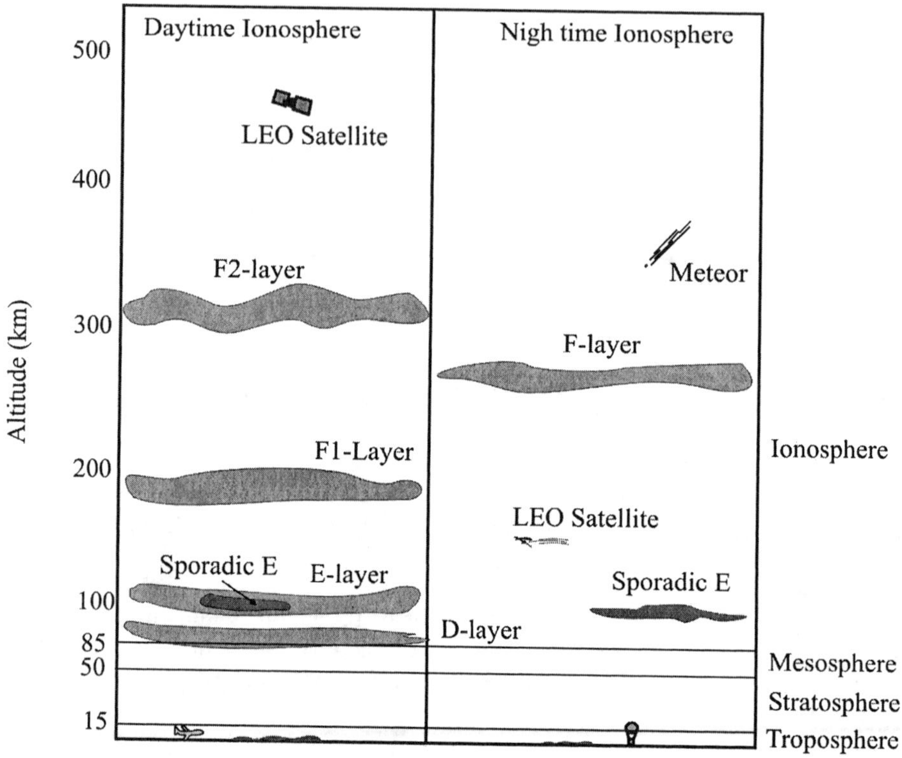

Figure 4.1 Structure of the atmosphere close to Earth.

The D-layer does not actually refract HF signals, but can attenuate them such that penetration is impossible. The E-layer refracts HF signals, as do the F-layers. Occasionally a highly ionized small region will traverse rapidly through the E-layer. This is called sporadic E, and it significantly changes the characteristics of the reflecting process. The frequency characteristics change as well as the amplitude of the reflection.

More than one reflection is possible by the EM wave reflecting off the ionosphere and returning to the ground, as shown in Figure 4.2. If the ground is suitably reflective (this is a reflection, not refraction), the signal will be reflected again upward. This can continue for several hops.

Short-term disturbances of the composition of the ionosphere in a region sometimes occur which impact HF radio propagation. They are super-ionized regions that last only a short time and are called *traveling ionospheric disturbances* (TID).

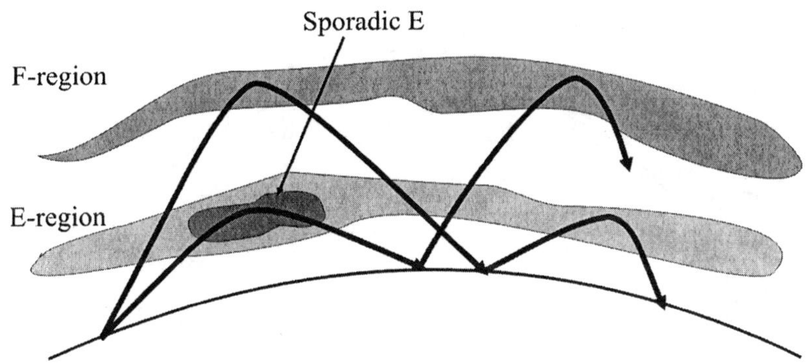

Figure 4.2 Ionospheric propagation modes vary, and can consist of several hops.

Short-range propagation of HF signals is also possible, of course. The modes of propagation involved with such signaling follow the same rules that *very high frequency* (VHF) and higher signals follow. The direct wave travels straight from the transmit antenna to the receive antenna. This can only occur, however, if the two antennas are within radio line-of-sight of one another.

There is also a phenomenon known as surface wave propagation that occurs much more predominantly in the HF range than in higher ranges. This is caused by the interaction of the Earth's characteristics (mainly the dielectric constant along the propagation route) with that of the air immediately above it. Such waves follow the curvature of the Earth and can propagate several hundreds of kilometers.

The geolocation of HF signals with these propagation modes is calculated by the same techniques as for VHF+ signals described elsewhere herein. The mode of propagation of interest in this chapter is that due to ionospheric reflections that, most of the time, are used for long-distance communications [*near vertical incidence sky wave* (NVIS) is an exception to this].

Figures 4.1 and 4.2 can be misleading because their dimensions were intentionally exaggerated to show detail. The actual HF signal propagation paths would be more like those shown in Figure 4.3.

4.1.1 Ionograms

The heights of the various ionospheric layers are measured with *sounders*. These are devices that send a signal upward and measure the time delay when a response is received. The frequency is varied (usually swept) so that the ionospheric heights are measured versus frequency. The resultant graph of virtual height versus frequency is called an *ionogram*.

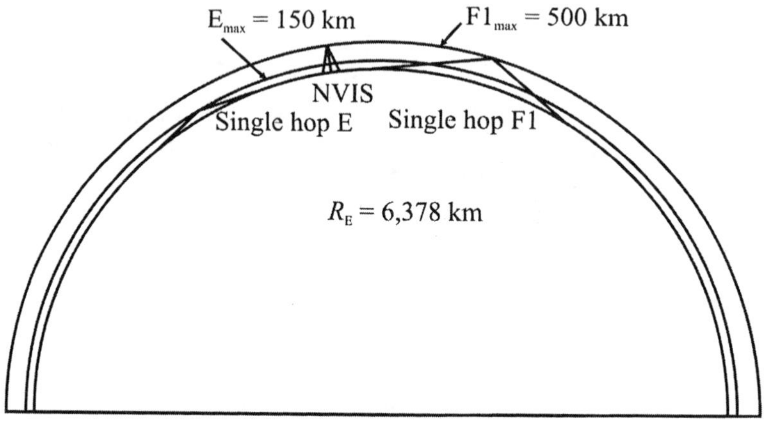

Figure 4.3 The Earth and ionosphere in correct perspective.

Vertical sounding is when the height is measured directly overhead of the receiving site. This is the only method typically available to SSL PF systems, since the targets to be located can normally be anywhere in azimuth and measuring the height of the ionosphere at other points would require many such sounders. When the ionospheric height is measured at some point other than directly overhead, it is referred to as *oblique sounding*. On HF communication paths such measurements are feasible, but, as mentioned, are not practical for calculating a PF.

The devices used to measure the height of the ionosphere versus frequency are called *ionospheric sounders*. Typically, a swept signal is radiated straight up for *vertical sounders* or at an angle for *oblique sounders*. The time of the return reflection is compared to when the signal was transmitted to ascertain the ionospheric height. The resultant displays of the sounder results are the ionograms. A typical vertical ionogram is illustrated in Figure 4.4 taken at Boulder, CO, on June 29, 1968, at 1930 hrs UT.[1]

Vertical sounders, of course, measure the ionospheric height directly overhead. It is frequently assumed that the ionosphere is homogeneous, and the heights are the same throughout the region. The height is usually measured at the site of the PF system and it is assumed to be the same height at the reflection point. This is rarely true, so such techniques for measuring the location of transmitters are not that accurate. EEPs with axes that are 10% of the range to the target are typical. Thus, if the target is 100 km from the PF system, the major axis

---

[1] *Universal time* (UT) is time at the Greenwich Observatory outside of London, UK, where the prime meridian (0° longitude) is located. UT is also known as *Greenwich time* (GT), *Greenwich mean time* (GMT), and Zulu time.

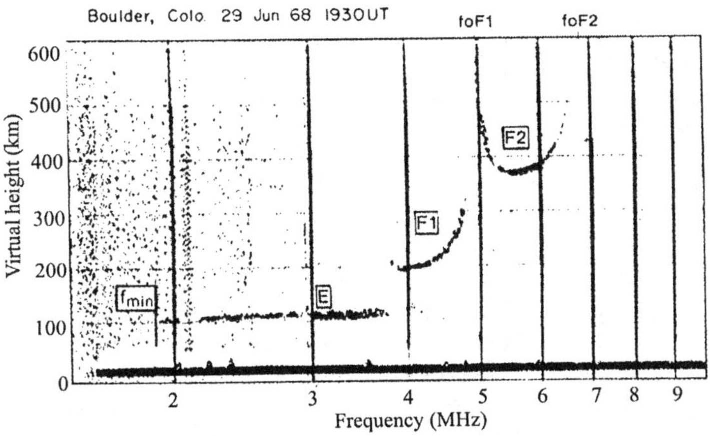

Figure 4.4 Typical ionogram display. This was obtained at Boulder, CO, on June 29, 1968 at 1930 hrs UT. The critical frequencies for the layers can be ascertained, as can the *lowest usable frequency* (LUF). The LUF is just below 2 MHz, while f_0E is just below 4 MHz, f_0F1 is approximately 5 MHz, and f_0F2 is about 6.6 MHz. (*Source:* [1], © 1989 Peter Peregrinus. Reprinted with permission.)

of the EEP can be 10 km or more. It should also be noted that although the reflecting surface discussed here was the ionosphere, any reflecting surface could also be used in some circumstances as long as the distance between that surface and the PF system is known. The effects of range between the emitter and receiver on the ionospheric characteristics are typified by the ionograms shown in Figure 4.5. The maximum frequency that can be used increases with increasing distance between the transmitter and receiver even though the critical frequency, which is always measured directly overhead, remains the same. In fact, the *maximum usable frequency* (MUF) and critical frequency are related by *Snell's law*:

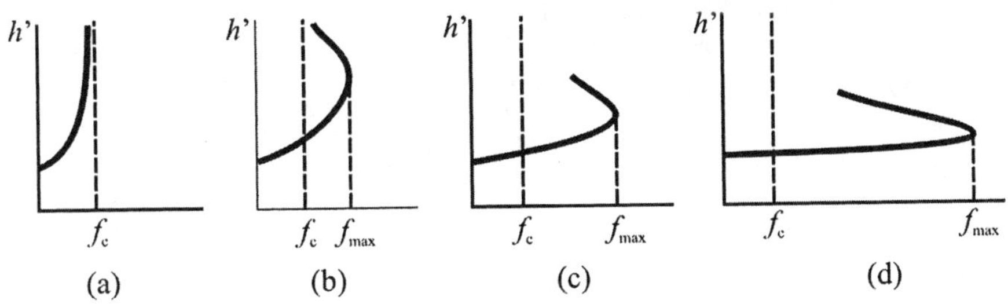

Figure 4.5 Oblique ionograms determined at different distances between the transmitter and receiver: (a) vertical, (b) short distance, (c) medium distance, and (d) long distance. Note that the trace is asymptotic to the critical frequency and that the maximum frequency increases with increasing distance. (*Source:* [2], © 1989 Peter Peregrinus. Reprinted with permission.)

$$f_{\max} = f_c \sec\theta_0 \tag{4.1}$$

where θ_0 is the zenith angle (measured from the virtual point of reflection). The critical frequency for the E-layer is denoted by f_0E, while that for F1 is denoted by f_0F1, and for F2 by f_0F2.

4.1.2 Magnetic Field Effects

The principal effect of the Earth's magnetic field on ionospheric waves is to split the wave into two: one being left-handed (with respect to the magnetic field), the so-called *ordinary wave*, and the other being right-handed, the so-called *extraordinary wave*. The directions here refer to the rotation direction of the polarization vector. These two components traverse the ionosphere via totally different paths. They recombine when they leave the ionosphere and the net effect is to change the polarization of the wave headed downward toward the receiver.

The frequencies of these two waves are different. Furthermore, the relationship between these frequencies depends on the orientation of the wave propagation relative to the orientation of the magnetic dipole of the Earth. Thus, waves propagating north-south or south-north will have different characteristics than those propagating east-west or west-east.

4.2 Single-Site Location

The source of HF signals propagating via sky wave paths can be located by triangulation using any of the techniques discussed previously. However, it is also possible to use a single PF system for this purpose under some circumstances. If the elevation of the ionospheric layer that the HF signal is refracting through or, more accurately, the equivalent height of the layer that is reflecting the signal is known, as shown in Figure 4.6, then the emitter can be located since we know the range to the target and its angle of arrival. The wave is assumed to be reflected at the midway point between the transmitter and the PF system. The elevation angle is related to the range and ionospheric height by

$$\tan\ \theta = \frac{h}{R/2} \tag{4.2}$$

so

$$R = \frac{2h}{\tan\ \theta} \tag{4.3}$$

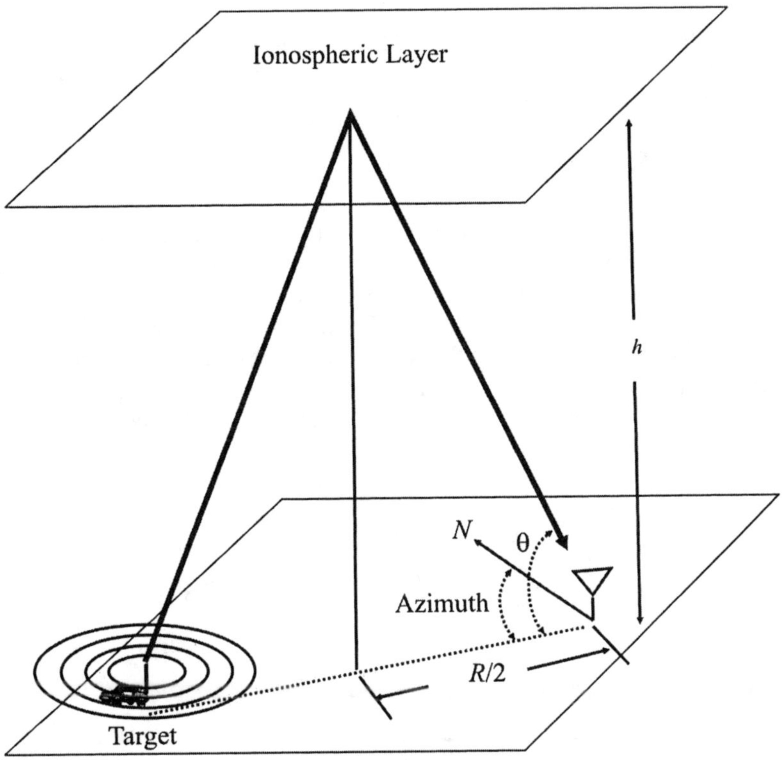

Figure 4.6 Single-site location of HF emitter.

Thus, by measuring the elevation *angle of arrival* (AOA), the range to the target can be estimated. Although this derivation was given using planes for the ionosphere and the Earth, it can also be derived using spherical surfaces, which is more accurate.

For the SSL technique to function properly, the number of times the signal has been reflected off the ionosphere and/or the ground must be known. Otherwise the range to the target cannot be determined. Signals that arrive from a longer range will arrive at a lower elevation angle and this may assist in determining the number of hops. Otherwise it is necessary to assume some number of reflections (usually one).

4.3 Passive SSL

When the signal arrives at the sensor from two or more directions, multipath propagation modes are present. In such cases it is possible to calculate the target location without sounding the height of the ionosphere. This is called *passive SSL.*

Consider the situation depicted in Figure 4.7 that shows the signal being reflected off two different ionospheric layers. According to the Briet and Tuve Theorem [3,4], even though the signals are in actuality refracted in the ionospheric layers, there are equivalent planes from which the signals are reflected. It is assumed that the Earth is flat, as are the equivalent reflecting planes, and the range can be determined as

$$\tan\theta = \frac{h}{R/2} \qquad \tan\phi = \frac{h+\Delta h}{R/2}$$

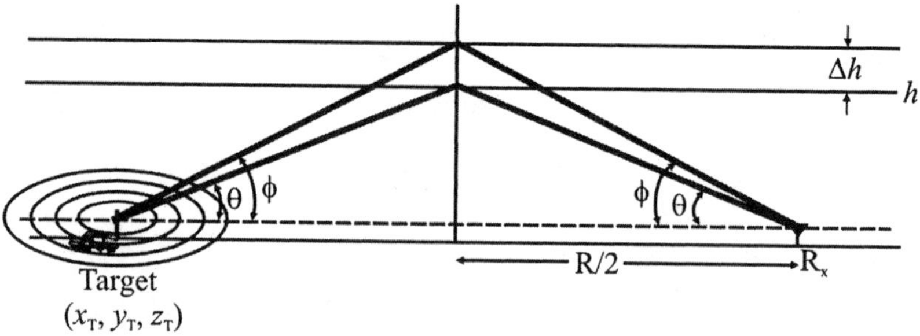

Figure 4.7 Multipath reflection.

$$h = \frac{R}{2}\tan\theta = \frac{R}{2}\tan\phi + \Delta h$$
$$R = \frac{2\Delta h}{\tan\phi - \tan\theta} \tag{4.4}$$

Now consider the detail of the signal and ionosphere interaction shown in Figure 4.8. The TDOA of the two signals at the receiver site is denoted by τ and the speed of propagation is denoted by c. Point A is the reflection point of the signal in the lower layer, while point C is that for the upper layer. Denoting the receive point as Rx (the transmit point could also be used with a mirror reflection of Figure 4.7), triangle ABR_x forms an isosceles triangle because the time to travel from A to R_x is the same as that from B to R_x, by construction of line AB. Summing the angles around the triangle defined by ABC then

$$\beta = \alpha + \phi - 90 \tag{4.5}$$

From triangle ABR_x,

$$2\alpha = 180 - (\phi - \theta)$$
$$\alpha = 90 - \frac{\phi - \theta}{2} \tag{4.6}$$

and

$$\angle ABC = 180 - \alpha$$
$$= 90 + \frac{\phi - \theta}{2} \tag{4.7}$$

From the law of sines,

$$\frac{c\tau/2}{\sin\left(\frac{\phi+\theta}{2}\right)} = \frac{\Delta h}{\sin\left(90 + \frac{\phi-\theta}{2}\right)} \tag{4.8}$$

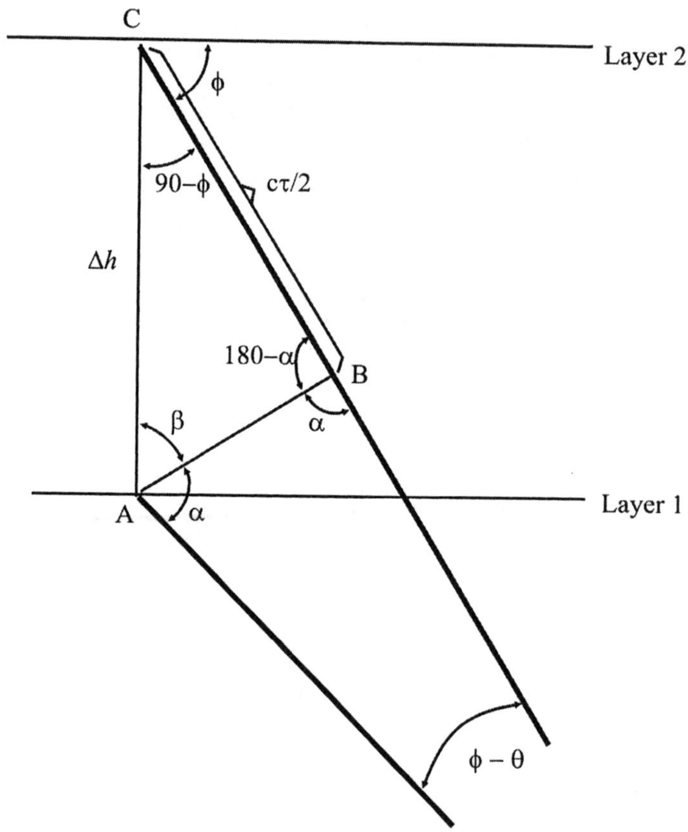

Figure 4.8 Multipath reflection details.

Solving (4.8) for Δh and substituting into (4.4), as well as using some trigonometric identities, yields

$$R = \frac{c\tau}{\tan\phi - \tan\theta} \frac{\cos\left(\frac{\phi-\theta}{2}\right)}{\sin\left(\frac{\phi+\theta}{2}\right)} \tag{4.9}$$

Therefore, the range can be determined by measuring the two elevation angles of arrival (if there are more than two, they can be combined two at a time in (4.9) and the result averaged).

As above, the target location is found by combining the range information with the azimuth angle of arrival, as illustrated in Figure 4.9.

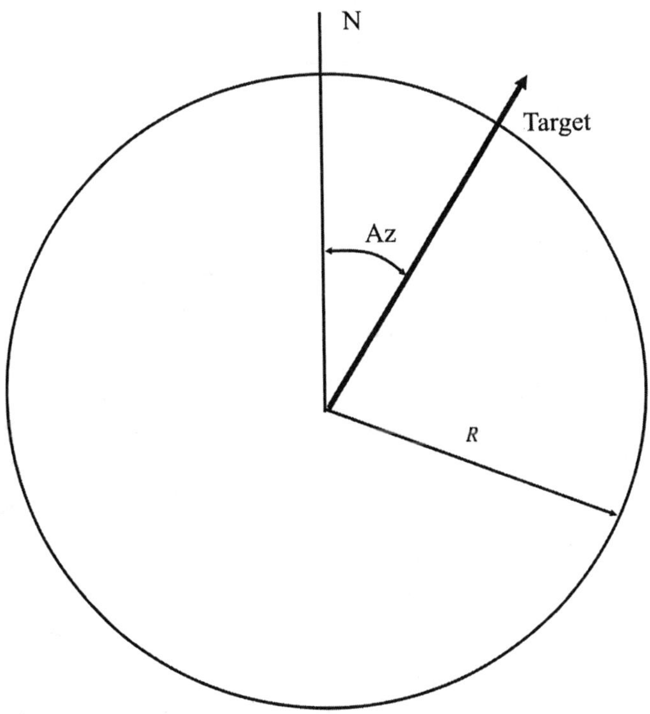

Figure 4.9 SSL target location.

4.4 Earth Curvature

The development above assumed that the surface of the Earth was a plane. For those cases where the range is 500 km or more, the curvature of the Earth cannot be ignored for purposes of determining the PF of the target. The geometry is illustrated in Figure 4.10. The distance from the receive sensor to the halfway point in the path is given by the arc s and

$$s = r\phi \tag{4.10}$$

when ϕ is measured in radians and r is the radius of the Earth (assumed to be spherical). The totality of the angles around triangle ABC is

$$\phi + \beta + \theta_{\text{EL}} = 90 \tag{4.11}$$

expressed in degrees. Finally, by the law of sines,

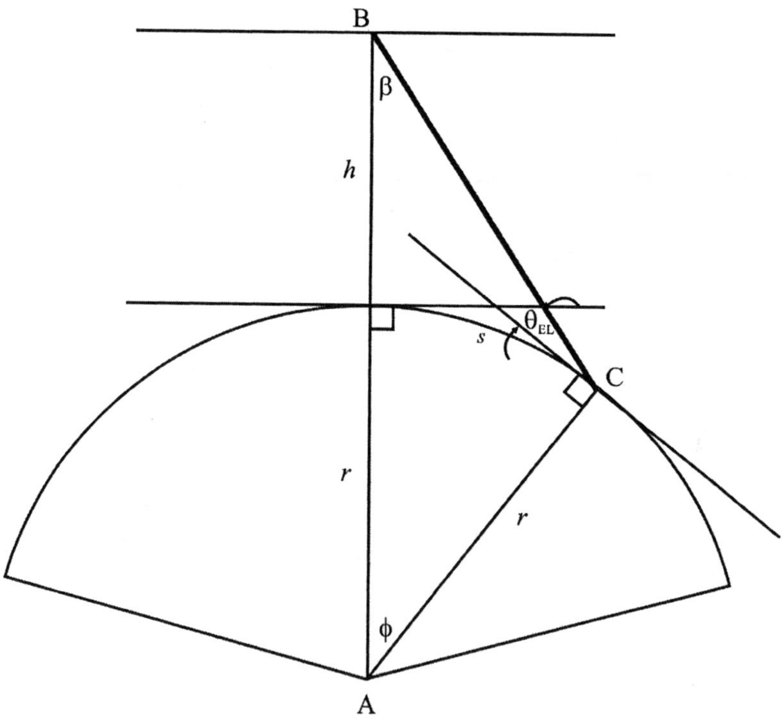

Figure 4.10 Effects of the curvature of the Earth's surface.

$$\frac{r}{\sin(90-\phi-\theta_{\text{EL}})}=\frac{r+h}{\sin(90+\theta_{\text{EL}})} \tag{4.12}$$

Manipulating (4.10) through (4.12) yields

$$s=r\tan^{-1}\left(\frac{\cos\theta_{\text{EL}}-\dfrac{r}{r+h}}{\sin\theta_{\text{EL}}}\right) \tag{4.13}$$

and $R = 2s$.

One of the effects of including the Earth's curvature is to decrease the MUF. This is because the effective distance between the transmitter and receiver is decreased.

The maximum one-hop range is determined by including the curvature. This range is given by

$$d_{\text{m}}=\sqrt{8rh_0} \tag{4.14}$$

and the MUF is given by

$$f_{\max}\approx f_{\text{c}}\sqrt{\left(\frac{r}{2h_0}\right)} \tag{4.15}$$

where h_0 is the reflection height.

4.5 Determining the TDOA with the Cepstrum

The *cepstrum* is one processing technique for determining the time delay between two versions of the same signal or the echoes of a signal that has been reflected. Knowledge of the time delay is necessary for measuring both the azimuth AOA and the elevation AOA for SSL PF estimation. Other methods for measuring differential time delay have also been proposed, including the *generalized correlation method* (GCM) [5] and the *adaptive eigenvalues decomposition algorithm* [6].

Its genesis lies in the technology of speech processing, but it has been widely applied outside of that area. Generally speaking, the complex cepstrum is found by computing the inverse *Fourier transform* (FT) of the logarithm of the FT of the signal [7]. The power cepstrum is also sometimes used. The complex cepstrum retains the phase information and therefore requires phase unwrapping, while the power cepstrum does not retain the phase.

If $c(t)$ represents the baseband impulse response of the propagation path and $g(t)$ represents the baseband impulse response of the transmitter and receiver, then the received demodulated signal is given by

$$r(t) = T(t) * h(t) * g(t) \tag{4.16}$$

where $T(t)$ is the source signal and * denotes convolution. The source signal represented at the receiver output is therefore given by

$$s(t) = g(t) * T(t) \tag{4.17}$$

Therefore,

$$r(t) = s(t) * c(t) \tag{4.18}$$

The goal is to determine $c(t)$ from (4.18). The power spectrum of $r(t)$ is given by

$$|R(f)| = |S(f)||C(f)| \tag{4.19}$$

Calculating the natural logarithm,

$$\ln|R(f)| = \ln|S(f)| + \ln|C(f)| \tag{4.20}$$

The inverse FT of (4.20) (denoted by *IFT*) yields the power cepstrum of $r(t)$

$$\begin{aligned} \hat{r}(t) &= IFT[\ln|R(f)|] \\ &= \hat{s}(t) + \hat{c}(t) \end{aligned} \tag{4.21}$$

As long as $\hat{s}(t)$ and $\hat{c}(t)$ do not overlap in the cepstrum domain, $\hat{c}(t)$ can be found by filtering.

Note that

$$FT\{IFT[\ln|R(f)|]\} = \ln|R(f)|$$
$$\exp\{\ln|R(f)|\} = |R(f)|$$

$$IFT|R(f)| = \gamma(t)$$

where $\gamma(t)$ is the autocorrelation function of $r(t)$.

To illustrate using the cepstrum for the purpose of determining the TDOA of the two multipath components, assume that the signal consists of impulse functions [8] that are by definition short, which are transmitted at intervals that are long enough so that the channel and all receiving equipment has settled by the time the next impulse occurs. In that case, the channel impulse response can be written as

$$r(t) = \delta(0) + A\delta(\tau) \tag{4.22}$$

where A is the normalized signal strength of the second multipath component and τ is the TDOA to be determined. The power spectrum of $r(t)$ is of the form

$$|R(f)| = 1 + A^2 + 2A\cos(2\pi f \tau) \tag{4.23}$$

which can be written in the form

$$|R(f)| = (1 + A^2)\left[1 + \frac{2A}{1 + A^2}\cos(2\pi f \tau)\right] \tag{4.24}$$

Now using the series expansion of ln $(1 + x)$ as

$$\ln(1 + x) = x - \frac{x^2}{2} + \frac{x^3}{3} - \frac{x^4}{4} + \cdots \qquad x < 1 \tag{4.25}$$

and ignoring all but the first term,

$$\ln(1 + x) \approx x \qquad x << 1 \tag{4.26}$$

Using this approximation in $\ln[|R(f)|]$ yields

$$\ln[|R(f)|] \approx \ln(1 + A^2) + 2A\cos(2\pi f \tau) \tag{4.27}$$

which is the cepstrum of $r(t)$. This cepstrum contains an impulse at the time delay equal to the TDOA.

The cepstrum is an example of an area of signal processing referred to as *homomorphic deconvolution* [9, 10]. Systems that obey the homomorphic rules are characterized by a generalization of the rules of superposition which are inherent to linear time invariant systems. Homomorphic deconvolution is a form of generalized filtering.

4.6 MUSIC Cepstrum SSL

Using (4.9) to compute the range to the target requires knowing the TDOA, τ, between the arrival times of the signals at two antennas at the receiver (typically, more than two antennas are used and the results combined). Johnson, Black, and Sonsteby developed a method of computing this TDOA when interference is present using an adaptation of the MUSIC algorithm [11].

The geometry considered is illustrated in Figure 4.11. The transmitted signal, $s(t)$, is reflected off a lower layer and a multipath version of the same signal is reflected off a higher layer. The signals received at the receiving array, denoted $r_1(t)$ and $r_2(t)$ corresponding to antenna 1 and antenna 2, respectively, are a combination of the lower signal and the multipath component. The received signals can be represented as

$$r_1(t) = A_g s(t) + A_s s(t-\tau) \tag{4.28}$$

$$r_2(t) = A_g s(t-\tau_g) + A_s s(t-\tau_s-\tau) \tag{4.29}$$

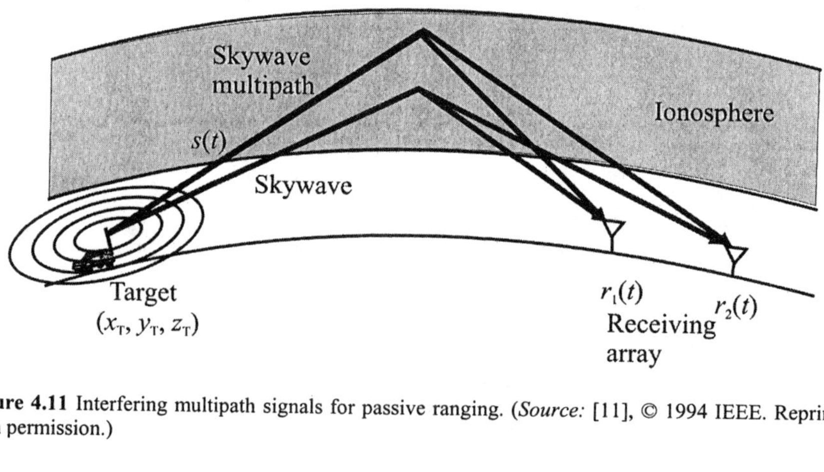

Figure 4.11 Interfering multipath signals for passive ranging. (*Source:* [11], © 1994 IEEE. Reprinted with permission.)

where τ_g is the propagation delay of the interfering signal between the two antennas at the receiver, τ_s is the propagation delay for the lower signal, and A_g is the path attenuation of the interfering signal, assumed the same for both receive antennas, since it is assumed that the antennas are much closer together than the height of the reflecting ionosphere. Likewise for A_s, which is the path attenuation of the lower signal. Calculating the Fourier transforms of (4.28) and (4.29) yields

$$\mathcal{F}[r_1(t)] = R_1(f) = (A_g + A_s e^{-j2\pi f\tau})S(f) \tag{4.30}$$

$$\mathcal{F}[r_2(t)] = R_2(f) = [A_g e^{-j2\pi f\tau_g} + A_s e^{-j2\pi f(\tau_g+\tau)}]S(f) \tag{4.31}$$

where $S(f)$ is the FT of $s(t)$.

The normalized cross-power spectrum of $r_1(t)$ and $r_2(t)$ is calculated to be

$$P_{12}(f) = \frac{R_1(f)R_2^*(f)}{\sqrt{|R_1(f)|^2\,|R_2(f)|^2}} \tag{4.32}$$

which, in polar form, becomes

$$P_{12}(f) = \frac{|A_1A_2^*|\arg(A_1A_2^*)|S(f)|^2}{\sqrt{A_1A_1^*\,|S(f)|^2\,A_2A_2^*\,|S(f)|^2}} \tag{4.33}$$

where

$$A_1 = A_g + A_s e^{-j2\pi f\tau} = |A_1|\angle A_1 \tag{4.34}$$

and

$$A_2 = A_g e^{-j2\pi\tau_g} + A_s e^{-j2\pi f(\tau_s+\tau)} = |A_2|\angle A_2 \tag{4.35}$$

Expression (4.33), after trigonometric manipulation and with the addition of noise, reduces to

$$P_{12}(f) = e^{j\phi(f)} + n(f) \tag{4.36}$$

where

$$\phi(f) = \tan^{-1}\left\{\frac{\begin{array}{l} A_g^2 \sin 2\pi f \tau_g + A_s^2 \sin 2\pi f \tau_s \\ \qquad + 2A_g A_s \cos\left[\pi f(2\tau + \tau_s - \tau_g)\right]\sin\left[\pi f(\tau_g + \tau_s)\right] \end{array}}{\begin{array}{l} A_g^2 \cos 2\pi f \tau_g + A_s^2 \cos 2\pi f \tau_s \\ \qquad + 2A_g A_s \cos\left[\pi f(2\tau + \tau_s - \tau_g)\right]\cos\left[\pi f(\tau_g + \tau_s)\right] \end{array}}\right\} \tag{4.37}$$

Assume that the signal is distributed over M frequencies so that

$$\mathbf{P}_{12}(f) = \begin{bmatrix} e^{j\phi(f_1)} \\ e^{j\phi(f_2)} \\ \vdots \\ e^{j\phi(f_M)} \end{bmatrix} + \begin{bmatrix} n(f_1) \\ n(f_2) \\ \vdots \\ n(f_M) \end{bmatrix} \tag{4.38}$$

The spectral cross-correlation matrix is given by

$$\mathbf{C}(f) = \mathcal{E}\{\mathbf{P}_{12}(f)\mathbf{P}_{12}^{\dagger}(f)\} \tag{4.39}$$

The ijth element of $\mathbf{C}(f)$ is given by

$$c_{ij}(f) = \mathcal{E}\{[e^{j\phi(f_i)} + n(f_i)][e^{j\phi(f_j)} + n(f_j)]\} \tag{4.40}$$

$$= e^{j[\phi(fi) - \phi(fj)]} + \sigma^2\delta(f_i - f_j) \tag{4.41}$$

It is the spectral cross-correlation matrix $\mathbf{C}$ upon which the MUSIC processing is performed.

MUSIC determines the eigen-decomposition of $\mathbf{C}$. This finds the eigenvalues and their associated eigenvectors. The eigenvalues $\{\lambda_i\}$ are ordered such that $\lambda_M \geq \lambda_{M-1} \geq \cdots \geq \lambda_1$ with corresponding eigenvector matrix

$$\mathbf{C}_{\mathrm{E}} = [\mathbf{E}_M \quad \mathbf{E}_{M-1} \quad \cdots \quad \mathbf{E}_1] \tag{4.42}$$

For a single pair of coherent signals, as is the case here, $\mathbf{C}_{\mathrm{E}}$ will have rank 1 and the nullspace matrix is given by the remaining eigenvectors as

$$\mathbf{C}_{\mathrm{N}} = [\mathbf{E}_{M-1} \quad \mathbf{E}_{M-1} \quad \cdots \quad \mathbf{E}_1] \tag{4.43}$$

The MUSIC spectrum is given as

$$P(f) = \frac{1}{\mathbf{B}^H \mathbf{C}_N \mathbf{C}_N^H \mathbf{B}} \tag{4.44}$$

where

$$\mathbf{B} = \begin{bmatrix} e^{j\phi(f_1)} & e^{j\phi(f_2)} & \cdots & e^{j\phi(f_M)} \end{bmatrix}^{\mathrm{T}} \tag{4.45}$$

To find the TDOA to use in (4.9) for calculation of the range, the spectrum given by (4.44) is searched for the peak value.

When there are p paths, as opposed to the two just discussed, then

$$\phi(f) = \tan^{-1}\left\{ \frac{\sum_{j=1}^{p}\sum_{i=1}^{p} k(i,j) V_i V_j \sin\left[\pi f(\tau_i + \tau_j)\right] \cos[\pi f(2\tau_{ij} - \tau_i + \tau_j)]}{\sum_{j=1}^{p}\sum_{i=1}^{p} k(i,j) V_i V_j \cos\left[\pi f(\tau_i + \tau_j)\right] \cos[\pi f(2\tau_{ij} - \tau_i + \tau_j)]} \right\} \tag{4.46}$$

$$k(i,j) = \begin{cases} 1, & i = j \\ 2, & i \neq j \end{cases} \tag{4.47}$$

$$\tau_{ii} = 0 \qquad \forall i \tag{4.48}$$

and the paths are characterized by

$$A_1 = \sum_{i=1}^{p} V_i e^{-j2\pi f \tau_{1i}} \tag{4.49}$$

and

$$A_2 = \sum_{i=1}^{p} V_i e^{-j2\pi f(\tau_{1i} + \tau_i)} \tag{4.50}$$

The amplitude of the ith path is V_i, the delay between antennas for path i is τ_i, and the delay between antenna i and the reference antenna (1) is τ_{i1}.

4.7 Ray Tracing

Ionospheric tilt can cause significant errors in PF computation with SSL techniques. If the ionosphere is not accurately modeled as a horizontal plane, the measured azimuth will be incorrect, as the effective tilt will cause the signal to be reflected at an incorrect angle.

The ionospheric tilts are caused by horizontal gradients of the refractive index in the ionosphere. This, in turn, is caused by nonuniform ionic densities. The effect is particularly problematic for SSL PF calculations when TIDs occur.

Mathematical models of the ionosphere have been developed in which it is not represented as a flat plane. *Ray tracing* is the appellation applied to utilization of these models, and, technically, the term also applies when the ionosphere is modeled as a plane. Theoretically, if details of the ionospheric structure are known, the ray paths taken by the signal between the transmitter and receiver can be retraced and the PF of the emitter can be determined by the intersection of the multiple paths. Measurement errors and noise, of course, prevent this from happening, and the characteristics of the ionosphere can only be determined to a certain degree of accuracy. Furthermore, in the case of PF on target emitters, the ionospheric height is normally measured at the location of the receiver, and a flat ionospheric plane is assumed. Measuring more details about the ionosphere is usually precluded when the targets in question are hostile. Therefore, modeling is a useful tool for increasing the accuracy of the measurement of the azimuth AOA.

The effects of refraction in a nonanisotropic medium can be accounted for using Fermat's principle of least path time.[2] Fermat's principle dictates that the path traversal time is minimized, which accounts for the apparent bending of a light ray at the interface of two media of different refractive indices. Using the calculus of variations, this can be expressed as

$$\delta \int_A^B \mu ds = 0 \tag{4.51}$$

where A and B are the two end points of the path and μ is the refractive index of the ionosphere. Converting this to polar coordinates,

$$ds = \sqrt{dr^2 + r^2 d\phi^2 + dz^2} \tag{4.52}$$

[2] An anisotropic medium is one that accounts for the magnetic-ionic interaction effects. The effects of the Earth's magnetic field can be taken into account but are ignored here.

where r is the distance from the center of the Earth, ϕ is the great circle distance angle, and z is the lateral distance. With

$$r' = \frac{dr}{d\phi} \qquad z' = \frac{dz}{d\phi} \qquad L = \mu\sqrt{r'^2 + r^2 + z'^2}$$

then, from (4.51),

$$\frac{d}{d\phi}\left(\frac{\partial L}{\partial r'}\right) - \frac{\partial L}{\partial r} = 0 \tag{4.53}$$

and

$$\frac{d}{d\phi}\left(\frac{\partial L}{\partial z'}\right) - \frac{\partial L}{\partial z} = 0 \tag{4.54}$$

Let

$$Q = \mu r'(r^2 + r'^2 + z'^2)^{-\frac{1}{2}} \tag{4.55}$$

and

$$W = \mu z'(r^2 + r'^2 + z'^2)^{-\frac{1}{2}} \tag{4.56}$$

Then, because for path P,

$$dP = \mu^{-1} ds \tag{4.57}$$

the following can be derived

$$\frac{dr}{dP} = Q \tag{4.58}$$

$$\frac{dQ}{dP} = \frac{1}{2}\frac{\partial \mu^2}{\partial r} + \frac{\mu^2 - Q^2 - W^2}{r} \tag{4.59}$$

$$\frac{d\phi}{dP} = \frac{(\mu^2 - Q^2 - W^2)^{1/2}}{r} \tag{4.60}$$

$$\frac{dW}{dP} = \frac{1}{2}\frac{\partial \mu^2}{\partial z} \tag{4.61}$$

$$\frac{dz}{dP} = W \tag{4.62}$$

Equations (4.58) through (4.62) specify the coordinates of a ray given in terms of r, z, and ϕ at every point along P given that the initial conditions and the refractive index are specified.

The refractive index is a function of the coordinates in the ionosphere and is approximated by

$$\mu(r,z,\phi) = \left[1 - \frac{\beta N(r,z,\phi)}{f^2}\right]^{1/2} \tag{4.63}$$

where $N(r,z,\phi)$ is the electron density (electrons per cubic centimeter), f is the frequency in megahertz (MHz), and $\beta = 8.05 \times 10^{-5}$ MHz^{2}cm^3.

Assuming that there are three layers in the ionosphere, the E, F1, and F2, then the electron density is a function of the height above the Earth, h, and is spherically symmetric. Using a cylindrical coordinate system, then, $N(r,z,\phi)$ is a function of r but not of z nor ϕ, (4.61) and (4.62) are not needed, and (4.58) to (4.60) reduce to a two-dimensional model.

The electron density is given by [12,13]

$$N(r_1) = N_0 e^{\frac{1}{2}[1 - r_1 - \sec\chi \exp(-r_1)]} \tag{4.64}$$

where χ is the zenith angle of the sun relative to the ionosphere, where it has been tacitly assumed that $dN(r)/dt$ is small. In addition,

$$r_1 = \frac{r - r_0}{H_\text{p}} \tag{4.65}$$

where r_0 is the reference height where the production of ions is maximum when the sun is directly overhead. H_p is the *pressure scale height* given by

$$H_p = \frac{KT}{\overline{m}g} \tag{4.66}$$

with

$\overline{m}$ = mean molelular mass

g = acceleration of gravity

T = temperature (Kelvin)

The irregular variations of the magnetic field disturbances are given by K.

When there are lateral gradients in the electron density in the ionosphere, then there is a dependence on altitude z and the electron density is denoted as $N(r,z)$. From (4.63)

$$\frac{\partial\mu(r,z)^2}{\partial r} = -\frac{\beta}{f^2}\frac{\partial N(r,z)}{\partial r} \tag{4.67}$$

and

$$\frac{\partial\mu(r,z)^2}{\partial z} = -\frac{\beta}{f^2}\frac{\partial N(r,z)}{\partial z} \tag{4.68}$$

For a ray that originates at sea level, then the initial conditions are that r is the approximated radius of the Earth, $Q = \sin\theta_{\text{EL}}$, and $W = z = 0$. This is sufficient information to solve (4.58) through (4.62) numerically.

The electron density in the ionospheric layers can be modeled as parabolas. In that case, several interesting characteristics emerge. An example of this is illustrated in Figure 4.12. In general, the entry and exit angles are not equal. They are, however, related by

$$\sin^2\phi_{\text{e}} - \sin^2\phi_{\text{i}} = A(x_{\text{e}} - x_{\text{i}}) \tag{4.69}$$

and

$$\cos^2\phi_{\text{e}} - \cos^2\phi_{\text{i}} = -C(z_{\text{e}}^2 - z_{\text{i}}^2) \tag{4.70}$$

where

$$A = -\frac{A_0}{f^2}$$

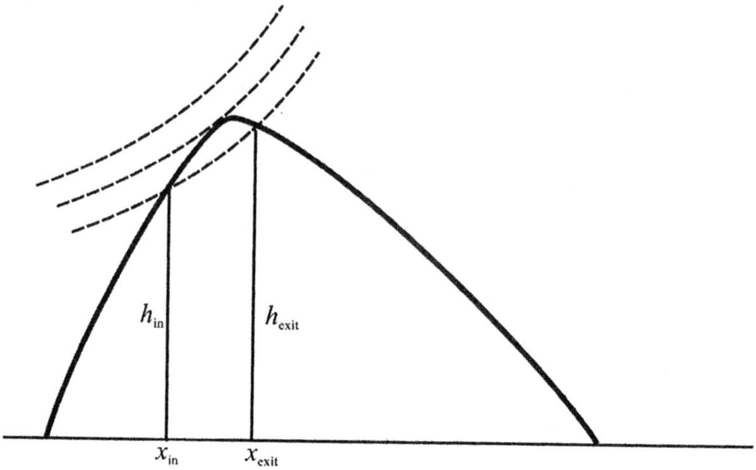

Figure 4.12 A horizontal tilt in the parabolic layer model causes an asymmetric ray path.

$$C = \left(\frac{f_c}{f y_m} \right)^2$$

A_0 is the gradient of $(f_0\text{F2})^2$ in the horizontal direction

$z = h_m - h$

y_m is the semithickness of the layer

h_m is the height of the maximum electron density

4.7.1 Parabolic Modeling

Parabolic mathematical models of the ionospheric layers have proven to be fairly reliable and they can be fit well to the layers with parameters determined from ionograms. Baker and Lambert [14] discuss multiquasiparabolic ionospheric models, initially developed by them [15] and Dyson and Bennett [16]. In that model, the electron density profile is given by

$$N(h) = N_m \left[1 \mp \left(\frac{h - h_m}{y_m} \right)^2 \right] \tag{4.71}$$

where N is the electron density at height h, N_m is the maximum electron density at height h_m, and y_m, as above, is the layer semithickness. The minus sign corresponds to a parabolic layer, while the plus sign applies to an inverted parabolic layer. An ionosphere with an E-layer and an F-layer is thus modeled with two normal parabolic layers with an inverted parabolic transition segment between them. The height h_{t2} at which the transition segment 2 blends with layer 3 and the semithickness of that transition segment y_{m2} are given by

$$h_{t2} = -\frac{\left[y_{m3}^2 \left(\frac{N_{m1}}{N_{m3}} - 1 \right) + h_{m3}^2 - h_{m1} h_{m3} \right]}{h_{m1} - h_{m3}} \tag{4.72}$$

and

$$y_{m2}^2 = \left(\frac{N_{m1}}{N_{m3}} \right) \left(-\frac{h_{t2} - h_{m1}}{h_{t2} - h_{m3}} \right) y_{m3}^2 \tag{4.73}$$

This model is illustrated in Figure 4.13 which compares two ionospheres with the Dudeney model. The two ionospheres are the same except the second one contains a slab of electrons of constant density above h_mE.

4.8 Concluding Remarks

Some methods for geolocating HF targets from a single sensor-site have been discussed in this chapter. While triangulation and quadratic processing methods apply to HF targets as well as higher frequency targets, in some cases it may be advantageous to apply only a single sensor. Such techniques typically result in lower accuracy geolocations because of the necessity of normally requiring knowledge of the height of the ionosphere. Measuring this height with EW systems in a noncooperative fashion is inherently inaccurate because of the assumption of uniform height of the reflecting layer above the Earth's surface.

In addition to their utility in EW applications, SSL EW systems have been applied to ionospheric research projects. Such projects endeavor to characterize the ionosphere, allowing for scientific study of the atmosphere.

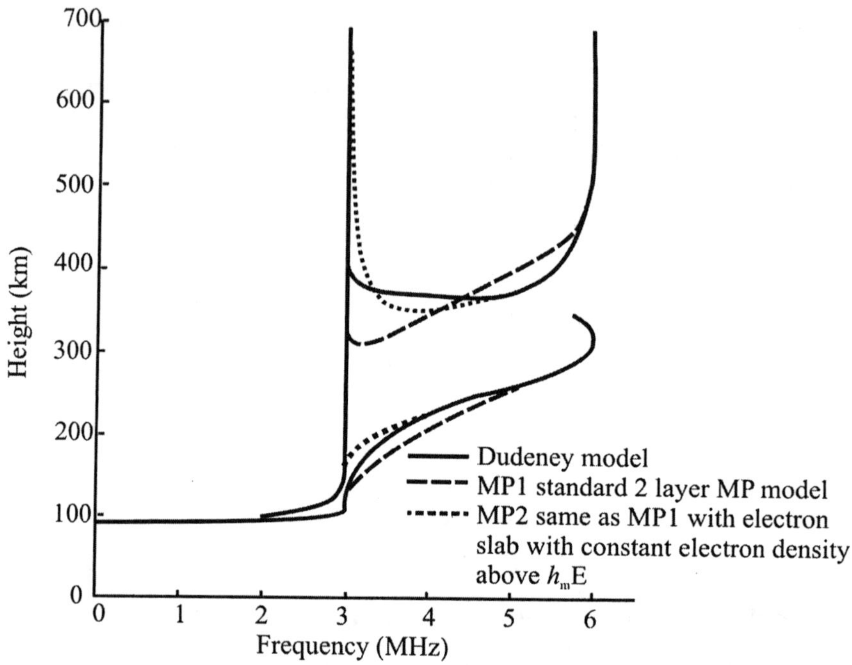

Figure 4.13 Plasma frequency profiles (lower curves) and corresponding ionograms (upper curves). In this case f_0E = 3 MHz, h_mE = 110 km, y_mE = 20 km, f_0F = 6 MHz, h_mF = 320 km, and y_mF = 100 km. (*Source:* [15], © 1988 IEE. Reprinted with permission.)

References

[1] Davies, K., *Ionospheric Radio,* London: Peter Peregrinus Ltd, 1989, p. 96.

[2] Davies, K., *Ionospheric Radio,* London: Peter Peregrinus Ltd, 1989, p. 164.

[3] Gething, P. J. D., *Radio Direction Finding and Superresolution,* 2nd Ed., London: Peter Peregrinus Ltd., 1978, pp. 262–265.

[4] Davies, K., *Ionospheric Radio,* London: Peter Peregrinus Ltd, 1989, p. 158.

[5] Knapp, C. H., and G. C. Carter, "The Generalized Correlation Method for Estimation of Time Delay," *IEEE Transactions on Acoustics, Speech, and Signal Processing*, Vol. ASSP-24, No. 4, August 1976, pp. 320–327.

[6] Benesty, J, "Adaptive Eigenvalue Decomposition Algorithm for Passive Acoustic Source Localization," *Journal of the Acoustic Society of America*, Vol. 107, No. 1, January 2000, pp. 384–391.

[7] Childers, D. G., D. P. Skinner, and R. C. Kemerait, "The Cepstrum: A Guide to Processing," *Proceedings of the IEEE,* Vol. 65, No. 10, October 1977, pp. 1428–1443.

[8] Eken, F., F. Atmaca, and E. Hepsaydir, "HF Multipath Dispersion Measurements Using Cepstral Signal Processing," *Radio Science,* Vol. 26, No. 1, Jan–Feb 1991, pp. 15–21.

[9] Roberts, L. R., *Signal Processing Techniques,* Anaheim, CA: Interstate Electronics Corporation, 1981, pp. 9-19–9-21.

[10] Oppenheim, A. V., and R. W. Schafer, *Digital Signal Processing,* Englewood Cliffs, NJ: Prentice Hall, 1975, Section 10.5.

[11] Johnson, R. L., Q. R. Black, and A. G. Sonsteby, "HF Multipath Passive Single Site Radio Location," *IEEE Transactions on Aerospace and Electronic Systems,* Vol. 30, No. 2, April 1994, pp. 462–470.

[12] Coleman, C. J., "On the Simulation of Backscatter Ionograms," *Journal of Atmospheric and Solar-Terrestrial Physics,* Vol. 59, 1997, pp. 2089–2099.

[13] Davies, K., *Ionospheric Radio,* London: Peter Peregrinus Ltd, 1989, p. 64.

[14] Baker, D. C., and S. Lambert, "Range Estimation for SSL HFDF Systems by Means of a Multi-quasiparabolic Ionospheric Model," *Proceedings of the IEE*, Vol. 136, Pt. H., No. 2, April 1989, pp. 120–125.

[15] Baker, D. C., and S. Lambert, "Multi-parabolic Ionospheric Model for SSL Application," *Electronic Letters,* 1988, 24, pp. 425–426.

[16] Dyson, P. L., and J. A. Bennett, "A Model of the Vertical Distribution of the Electron Concentration in the Ionosphere and its Application to Oblique Propagation Studies," *Journal of Atmospheric Terrestrial Physics,* 1988, 50, pp. 251–262.

Appendix A

Grassmann Algebra

A.1 Background

Hermann Gunther Grassmann was born in 1809 near the border of Germany and Poland in a town called Stettin. He discovered an algebra of geometry that includes vectors. His algebra subsumes the common form of algebra we all learn in high school. During his lifetime he received very little recognition for his discovery. William Kingdon discovered Grassmann's work about 1877, the year Grassmann died. Clifford's algebras are still used today for manipulation of vectors and linear mathematical systems.

A brief introduction to Grassmann algebra is presented in this appendix. The information included here is largely based on reference [1]. For those desiring more detailed information references [2–21] are recommended.

A.2 Grassmann Algebra

Grassmann's algebra is a theory of "extension," with the fundamental product operation being the exterior product. In this algebra, addition, negation, and multiplication by a scalar carry the same meaning as in the normal mathematics we are all familiar with. The product operation, however, is different.

The constructs of this algebra allow manipulation of geometric objects algebraically. That is, lines can be constructed from points, and lines can be manipulated. Planes can be constructed from lines and/or points and these planes can be manipulated algebraically. Higher forms of geometric shapes can be constructed and manipulated just as easily. Such manipulation includes functions such as finding the intersection of two or more geometric objects.

The common 3-D vectors are one form of geometric shapes that can be built in the algebra. Vectors such as these, of course, have a direction as well as magnitude. However, that is not enough to completely specify a vector. It is also necessary to know where the vector is located in the 3-D space to determine the effects. For example, if the vector represents a force exerted on an object, it is necessary to know where on the object the force is exerted, yet there is no information inherent in the specification of the vector to convey this information. Grassmann algebra solves this problem as part of the specification of the vector. All the necessary information about the force is part of the vector.

The concept of direction associated with vectors is included as part of the specification of planes and higher order objects. Thus a plane has a "direction." On the other hand, the concept of magnitude associated with vectors is not necessary for the manipulation of geometric objects, although it can be included if desired. This lack of magnitude is manifest in the lack of necessity of specifying a metric in the algebra, which is necessary in the familiar algebra of vectors.

A.3 Exterior Product

Grassmann algebra is a mathematical system that describes and manipulates physical entities. It is a linear or vector algebra such that the product operation facilitates the notion of linear dependence. The product operation is normally called the *exterior product* or *wedge product* and is denoted by $\wedge$. Linear dependence is encapsulated in the notion that if vectors $\mathbf{x}_1$, $\mathbf{x}_2$, … are linearly dependent, then their exterior product is zero:

$$\mathbf{x}_1 \wedge \mathbf{x}_2 \wedge \ldots \wedge \mathbf{x}_n = 0 \tag{A.1}$$

On the other hand, if the vectors are linearly independent, then their exterior product is nonzero.

A vector $\mathbf{x}_1$ can be interpreted as a direction in a linear space. If that space has a metric,[1] the $|\mathbf{x}_1|$, the magnitude of $\mathbf{x}_1$, is its length. The exterior product extends this vector to higher dimensions. The wedge product of two vectors, $\mathbf{x}_1 \wedge \mathbf{x}_2$, is called the *bivector* and can be thought of as the analog of direction to two-space, that is, the *planar direction*. If this two-space has a metric, then the magnitude of $\mathbf{x}_1 \wedge \mathbf{x}_2$ is its area. This can be extended to any number of dimensions.

Lines and planes are but examples of higher order constructs that originate with points. A line is represented by the exterior product of any two points on it, or

[1] A metric is a generalization of distance. The *metric* for a space consisting of vectors $\mathbf{x}_1$, $\mathbf{x}_2$, and so on, is a function d such that

(1) $d(\mathbf{x}_1, \mathbf{x}_2) = 0$ if and only if $\mathbf{x}_1 = \mathbf{x}_2$;

(2) $d(\mathbf{x}_1, \mathbf{x}_2) = d(\mathbf{x}_2, \mathbf{x}_1)$;

(3) $d(\mathbf{x}_1, \mathbf{x}_2) \le d(\mathbf{x}_1, \mathbf{x}_3) + d(\mathbf{x}_3, \mathbf{x}_2)$.

by any point on it with a vector parallel to it. A plane is the exterior product of any three points on it, any two points with a vector parallel to it, or any point on it with a trivector parallel to it.

The real algebra familiar to all of us as learned in high school is subsumed by Grassmann algebra. In that case the exterior product reduces to the normal product.

A.3.1 Properties of the Exterior Product

Given the exterior product of two vectors $\mathbf{x}_1$ and $\mathbf{x}_2$, $\mathbf{x}_1 \wedge \mathbf{x}_2$ is the bivector corresponding to them. This bivector is not a vector and therefore does not belong to the vector space in which $\mathbf{x}_1$ and $\mathbf{x}_2$ lie. The set of bivectors constructed with the vectors in the vector space in fact forms its own linear space.

The exterior product exhibits *antisymmetry* (or anticommutativity). That is,

$$\mathbf{x}_1 \wedge \mathbf{x}_2 = -\mathbf{x}_2 \wedge \mathbf{x}_1 \tag{A.2}$$

Therefore,

$$\begin{aligned} \mathbf{x}_1 \wedge \mathbf{x}_1 &= -\mathbf{x}_1 \wedge \mathbf{x}_1 \\ \mathbf{x}_1 \wedge \mathbf{x}_1 + \mathbf{x}_1 \wedge \mathbf{x}_1 &= 0 \\ 2\mathbf{x}_1 \wedge \mathbf{x}_1 &= 0 \\ \mathbf{x}_1 \wedge \mathbf{x}_1 &= 0 \end{aligned} \tag{A.3}$$

which indicates that $\mathbf{x}_1$ is linearly dependent on itself (the coefficient is 1; that is, $\mathbf{x}_1 = 1\mathbf{x}_1$).

Properties

(1) The exterior product is associative, $(\mathbf{x}_1 \wedge \mathbf{x}_2) \wedge \mathbf{x}_3 = \mathbf{x}_1 \wedge (\mathbf{x}_2 \wedge \mathbf{x}_3)$

(2) The exterior product is distributive, $\mathbf{x}_1 \wedge (\mathbf{x}_2 + \mathbf{x}_3) = \mathbf{x}_1 \wedge \mathbf{x}_2 + \mathbf{x}_1 \wedge \mathbf{x}_3$

(3) If a and b are any scalars, then

$$\begin{aligned} (a+b) \wedge \mathbf{x}_1 &= a\mathbf{x}_1 \wedge b\mathbf{x}_1 \\ a\mathbf{x}_1 \wedge \mathbf{x}_2 &= a(\mathbf{x}_1 \wedge \mathbf{x}_2) \end{aligned}$$

(4) Let $\mathbf{e}_1$, $\mathbf{e}_2$, and $\mathbf{e}_3$ be three basis vectors in 3-D space. Then any vectors can be expressed in terms of these vectors as

$$\mathbf{x}_1 = a_1\mathbf{e}_1 + a_2\mathbf{e}_2 + a_3\mathbf{e}_3 \tag{A.4}$$

$$\mathbf{x}_2 = b_1\mathbf{e}_1 + b_2\mathbf{e}_2 + b_3\mathbf{e}_3 \tag{A.5}$$

where a_i and b_i are appropriate scalars. The bivector formed from $\mathbf{x}_1$ and $\mathbf{x}_2$ is given as

$$\begin{aligned}
\mathbf{x}_1 \wedge \mathbf{x}_2 &= (a_1\mathbf{e}_1 + a_2\mathbf{e}_2 + a_3\mathbf{e}_3) \wedge (b_1\mathbf{e}_1 + b_2\mathbf{e}_2 + b_3\mathbf{e}_3) \\
&= a_1\mathbf{e}_1 \wedge b_1\mathbf{e}_1 + a_1\mathbf{e}_1 \wedge b_2\mathbf{e}_2 + a_1\mathbf{e}_1 \wedge b_3\mathbf{e}_3 \\
&\qquad + a_2\mathbf{e}_2 \wedge b_1\mathbf{e}_1 + a_2\mathbf{e}_2 \wedge b_2\mathbf{e}_2 + a_2\mathbf{e}_2 \wedge b_3\mathbf{e}_3 \\
&\qquad + a_3\mathbf{e}_3 \wedge b_1\mathbf{e}_1 + a_3\mathbf{e}_3 \wedge b_2\mathbf{e}_2 + a_3\mathbf{e}_3 \wedge b_3\mathbf{e}_3 \\
&= a_1b_1\mathbf{e}_1 \wedge \mathbf{e}_1 + a_1b_2\mathbf{e}_1 \wedge \mathbf{e}_2 + a_1b_3\mathbf{e}_1 \wedge \mathbf{e}_3 \\
&\qquad + a_2b_1\mathbf{e}_2 \wedge \mathbf{e}_1 + a_2b_2\mathbf{e}_2 \wedge \mathbf{e}_2 + a_2b_3\mathbf{e}_2 \wedge \mathbf{e}_3 \\
&\qquad + a_3b_1\mathbf{e}_3 \wedge \mathbf{e}_1 + a_3b_2\mathbf{e}_3 \wedge \mathbf{e}_2 + a_3b_3\mathbf{e}_3 \wedge \mathbf{e}_3 \\
&= (a_1b_2 - a_2b_1)\mathbf{e}_1 \wedge \mathbf{e}_2 + (a_2b_3 - a_3b_2)\mathbf{e}_2 \wedge \mathbf{e}_3 + (a_3b_1 - a_1b_3)\mathbf{e}_3 \wedge \mathbf{e}_1
\end{aligned} \tag{A.6}$$

The coefficients in (A.6) are those obtained by the normal cross product of $\mathbf{x}_1$ and $\mathbf{x}_2$. However, the bivector space does not require a metric. The cross product generates a vector orthogonal to $\mathbf{x}_1$ and $\mathbf{x}_2$ and therefore assumes the existence of a metric. In addition, the cross product applies to two vectors in 3-D space. The exterior product formulation presented here applies to vector spaces of any, arbitrary dimension. For example, the trivector derived just as (A.6) was derived yields

$$\mathbf{x}_1 \wedge \mathbf{x}_2 \wedge \mathbf{x}_3 = (a_1b_2c_3 - a_3b_2c_1 + a_2b_3c_1 + a_3b_1c_2 - a_2b_3c_2 - a_2b_1c_3)\mathbf{e}_1 \wedge \mathbf{e}_2 \wedge \mathbf{e}_3 \tag{A.7}$$

(5) A trivector such as (A.7) has a single component $(\mathbf{e}_1 \wedge \mathbf{e}_2 \wedge \mathbf{e}_3)$ whose coefficient is the determinant of the constituent vectors given by the term $(a_1b_2c_3 - a_3b_2c_1 + a_2b_3c_1 + a_3b_1c_2 - a_2b_3c_2 - a_2b_1c_3)$. In any metric space, such as considered here, this coefficient is the volume of the 3-D parallelepiped formed from the constituent vectors. Note that if x_1, x_2, and x_3 are lying in a plane, then they are dependent and the trivector is zero, indicating that the volume is zero.

A.3.2 *m*-Vectors

The exterior product of m 1-vectors is called an m-*vector*. The number m is referred to as the *grade* of the m-vector. It is common practice to annotate an m-vector with its grade as an underscript. Thus, $\mathbf{x}_1 \wedge \mathbf{x}_2 \wedge \mathbf{x}_3 \wedge \mathbf{x}_4$ is of grade 4, which is denoted $\underset{4}{\boldsymbol{\gamma}} = \mathbf{x}_1 \wedge \mathbf{x}_2 \wedge \mathbf{x}_3 \wedge \mathbf{x}_4$. A 1-vector is usually not endowed with the underscript. The grade of a scalar is zero. With the dimension of the linear space denoted by *N*, the complementary grade of an *m*-vector is $N - m$.

Property: Interchanging Order. Interchanging the order of any factors in the exterior product, adjacent or otherwise, changes the sign of the product:

$$\ldots \wedge \mathbf{x}_i \wedge \ldots \mathbf{x}_j \ldots = -(\ldots \wedge \mathbf{x}_j \wedge \ldots \mathbf{x}_i \ldots) \tag{A.8}$$

Property: Grade of *m*-Vectors. The exterior product of an *m*-vector $\underset{m}{\boldsymbol{\alpha}}$ and *k*-vector $\underset{k}{\boldsymbol{\beta}}$ is an $(m + k)$ vector:

$$\underset{m}{\boldsymbol{\alpha}} \wedge \underset{k}{\boldsymbol{\beta}} = \underset{m+k}{\boldsymbol{\gamma}} \tag{A.9}$$

Property: Identity Element. The *identity* for the exterior product is the unit scalar:

$$\underset{m}{\boldsymbol{\alpha}} = 1 \wedge \underset{m}{\boldsymbol{\alpha}} = \underset{m}{\boldsymbol{\alpha}} \wedge 1 \tag{A.10}$$

Property: Scalar Factors. Scalars factor out of products

$$\left(a\,\underset{m}{\boldsymbol{\alpha}}\right) \wedge \underset{k}{\boldsymbol{\beta}} = \underset{m}{\boldsymbol{\alpha}} \wedge \left(a\,\underset{k}{\boldsymbol{\beta}}\right) = a\left(\underset{m}{\boldsymbol{\alpha}} \wedge \underset{k}{\boldsymbol{\beta}}\right) \tag{A.11}$$

Property: AntiCommutative. An exterior product is anticommutative whenever the grades of both vectors is odd:

$$\underset{m}{\boldsymbol{\alpha}} \wedge \underset{k}{\boldsymbol{\beta}} = (-1)^{mk}\, \underset{k}{\boldsymbol{\beta}} \wedge \underset{m}{\boldsymbol{\alpha}} \tag{A.12}$$

It is only vectors of odd grade that are antisymmetric. If at least one of the vectors of an exterior product has even grade, then the product commutes—that is, it is symmetric:

$$(\mathbf{x}_1 \wedge \mathbf{x}_2) \wedge \mathbf{x}_3 = \mathbf{x}_3 \wedge (\mathbf{x}_1 \wedge \mathbf{x}_2) \tag{A.13}$$

Property: Distributivity. The exterior product is both left and right distributive under addition:

$$\left(\underset{m}{\alpha}+\underset{m}{\beta}\right)\wedge\underset{k}{\gamma}=\underset{m}{\alpha}\wedge\underset{k}{\gamma}+\underset{m}{\beta}\wedge\underset{k}{\gamma} \qquad \underset{m}{\alpha}\wedge\left(\underset{k}{\beta}+\underset{k}{\gamma}\right)=\underset{m}{\alpha}\wedge\underset{k}{\beta}+\underset{m}{\alpha}\wedge\underset{k}{\gamma} \tag{A.14}$$

Property: Associativity. The exterior product is *associative*:

$$\left(\underset{m}{\alpha}\wedge\underset{k}{\beta}\right)\wedge\underset{n}{\gamma}=\underset{m}{\alpha}\wedge\left(\underset{k}{\beta}\wedge\underset{n}{\gamma}\right) \tag{A.15}$$

A.4 Regressive Product

The regressive product can be thought of as a *dual product* to the exterior product. It is used to find the intersection of geometrical objects. The regressive product is denoted by $\vee$.

A.4.1 Unions and Intersections of Spaces

Any given 1-element **z** is in the subspace spanned by **x** and **y** if the exterior product of (nonzero) $\mathbf{x} \wedge \mathbf{y}$ with **z** is zero. Thus, $\mathbf{x} \wedge \mathbf{y}$ is an element that can be used to *define* the subspace instead of the individual 1-elements **x** and **y**.

Thus, the *space* of $\mathbf{x} \wedge \mathbf{y}$ as the set of all 1-vectors **z** such that $\mathbf{x} \wedge \mathbf{y} \wedge \mathbf{z} = \mathbf{0}$. This can be extended to more general elements by defining the space of a simple element $\underset{m}{\alpha}$ as the set of all 1-elements β such that $\underset{m}{\alpha}\wedge\beta=0$.

Two elements are *congruent* if one is a scalar factor times the other. In a 3-D vector space, this is equivalent to saying that two vectors are congruent if they point in the same direction. The two vectors need not have the same magnitude nor the same point of origin. In fact, for general linear spaces, magnitude as a concept may not be necessary—it only becomes necessary when the space has a metric defined. The concepts of union and intersection only make sense up to congruence. Congruent elements define the same subspace.

A *union* of elements is an element defining the subspace they span *together*. It is similar in concept to the common union of sets. If A and B denote two not necessarily distinct sets of objects, then the union of A and B, usually denoted by $A \cup B$, is the set comprised of elements from A or elements from B. An *intersection* of elements is an element defining the subspace they span *in common*. Again, from the concepts of sets, the intersection of A and B, denoted by $A \cap B$, is defined as the set comprised of elements from A and B. See Figure A.1.

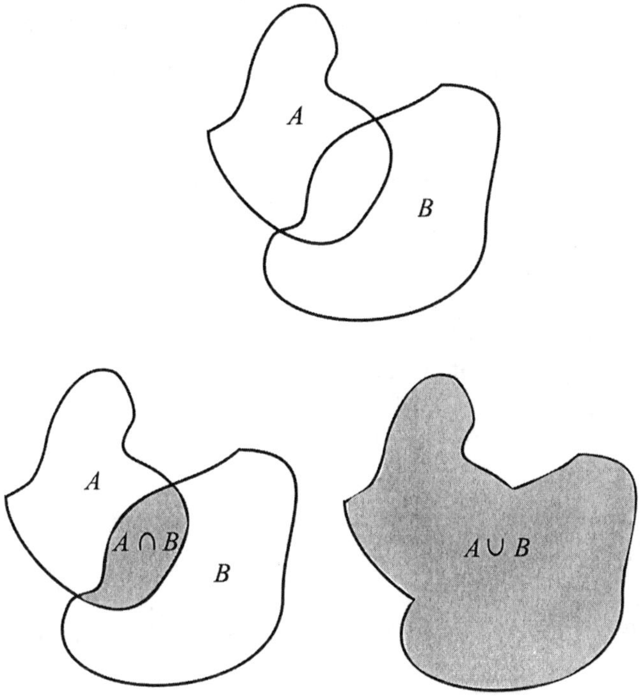

Figure A.1 Union and intersection of sets.

Given three independent 1-elements, **x**, **y**, and **z**, then a union of $\mathbf{x} \wedge \mathbf{y}$ and $\mathbf{y} \wedge \mathbf{z}$ is any element congruent to $\mathbf{x} \wedge \mathbf{y} \wedge \mathbf{z}$. An intersection of $\mathbf{x} \wedge \mathbf{y}$ and $\mathbf{y} \wedge \mathbf{z}$ is any element congruent to **y**.

A.4.2 Properties of the Regressive Product

The regressive product of an *m*-element and a *k*-element in an *N*-space is an $(m + k - N)$-element.

Property: Associative. The regressive product is associative:

$$\left(\underset{m}{\boldsymbol{\alpha}} \vee \underset{k}{\boldsymbol{\beta}}\right) \vee \underset{n}{\boldsymbol{\gamma}} = \underset{m}{\boldsymbol{\alpha}} \vee \left(\underset{k}{\boldsymbol{\beta}} \vee \underset{n}{\boldsymbol{\gamma}}\right) \tag{A.16}$$

Propety: Unit Element. The unit *n*-element $\underset{N}{\mathbf{1}}$ is the identity under the regressive product:

$$\underset{m}{\boldsymbol{\alpha}} = \underset{N}{\mathbf{1}} \vee \underset{m}{\boldsymbol{\alpha}} = \underset{m}{\boldsymbol{\alpha}} \vee \underset{N}{\mathbf{1}} \tag{A.17}$$

Property: Scalar Factors. Scalars factor out of products.

$$\left(a\underset{m}{\boldsymbol{\alpha}}\right) \vee \underset{k}{\boldsymbol{\beta}} = \underset{m}{\boldsymbol{\alpha}} \vee \left(a\underset{k}{\boldsymbol{\beta}}\right) = a\left(\underset{m}{\boldsymbol{\alpha}} \vee \underset{k}{\boldsymbol{\beta}}\right) \tag{A.18}$$

Property: Anticommutative. A regressive product is anticommutative whenever the complementary grades of the factors are both odd:

$$\underset{m}{\boldsymbol{\alpha}} \vee \underset{k}{\boldsymbol{\beta}} = (-1)^{(N-m)(N-k)} \underset{k}{\boldsymbol{\beta}} \vee \underset{m}{\boldsymbol{\alpha}} \tag{A.19}$$

Property: Distributive. The regressive product is both left and right distributive under addition:

$$\left(\underset{m}{\boldsymbol{\alpha}} + \underset{m}{\boldsymbol{\beta}}\right) \vee \underset{k}{\boldsymbol{\gamma}} = \underset{m}{\boldsymbol{\alpha}} \vee \underset{k}{\boldsymbol{\gamma}} + \underset{m}{\boldsymbol{\beta}} \vee \underset{k}{\boldsymbol{\gamma}} \qquad \underset{m}{\boldsymbol{\alpha}} \vee \left(\underset{k}{\boldsymbol{\beta}} + \underset{k}{\boldsymbol{\gamma}}\right) = \underset{m}{\boldsymbol{\alpha}} \vee \underset{k}{\boldsymbol{\beta}} + \underset{m}{\boldsymbol{\alpha}} \vee \underset{k}{\boldsymbol{\gamma}} \tag{A.20}$$

A.4.3 The Common Factor Axiom

Assume that **x**, **y**, and **z** are three independent vectors in a vector 3-space. The *common factor axiom* states that the regressive product of the two bivectors $\mathbf{x} \wedge \mathbf{z}$ and $\mathbf{y} \wedge \mathbf{z}$ may also be expressed as the regressive product of a 3-element $\mathbf{x} \wedge \mathbf{y} \wedge \mathbf{z}$ with **z:**

$$(\mathbf{x} \wedge \mathbf{z}) \vee (\mathbf{y} \wedge \mathbf{z}) = (\mathbf{x} \wedge \mathbf{y} \wedge \mathbf{z}) \vee \mathbf{z} \tag{A.21}$$

Since the space is 3-D, we can write any 3-element such as $\mathbf{x} \wedge \mathbf{y} \wedge \mathbf{z}$ as a scalar factor times the unit 3-element:

$$(\mathbf{x} \wedge \mathbf{y} \wedge \mathbf{z}) \vee \mathbf{z} = \left(a\underset{3}{\mathbf{1}}\right) \vee \mathbf{z} = a\mathbf{z} \tag{A.22}$$

Thus, the regressive product of two elements possessing an element in common is congruent to that element. That is, the regressive product of two elements defines their intersection:

$$(\mathbf{x} \wedge \mathbf{z}) \vee (\mathbf{y} \wedge \mathbf{z}) \equiv \mathbf{z} \tag{A.23}$$

Let $\underset{m}{\boldsymbol{\alpha}}, \underset{k}{\boldsymbol{\beta}}, \text{and} \underset{n}{\boldsymbol{\gamma}}$ be *simple* elements with $m + k + n = N$, where N is the dimension of the space. Then the common factor axiom states that

$$\left(\underset{m}{\boldsymbol{\alpha}} \wedge \underset{n}{\boldsymbol{\gamma}}\right) \vee \left(\underset{k}{\boldsymbol{\beta}} \wedge \underset{n}{\boldsymbol{\gamma}}\right) = \left(\underset{m}{\boldsymbol{\alpha}} \wedge \underset{k}{\boldsymbol{\beta}} \wedge \underset{n}{\boldsymbol{\gamma}}\right) \vee \underset{n}{\boldsymbol{\gamma}} \qquad m + k + n = N \tag{A.24}$$

When n is zero, the common factor axiom shows that the regressive product of an m-element with an $(n - m)$-element is a scalar that can be expressed as a regressive product with the unit **1**.

$$\underset{m}{\boldsymbol{\alpha}} \vee \underset{k}{\boldsymbol{\beta}} = \left(\underset{m}{\boldsymbol{\alpha}} \wedge \underset{N-m}{\boldsymbol{\beta}}\right) \vee \mathbf{1}$$

A.4.4 The Common Factor Theorem

Suppose that $\mathbf{x} \wedge \mathbf{y}$ and $\mathbf{u} \wedge \mathbf{v}$ are noncongruent bivectors in a 3-D space. The regressive product of $\mathbf{x} \wedge \mathbf{y}$ and $\mathbf{u} \wedge \mathbf{v}$ is given by

$$\mathbf{z} = (\mathbf{x} \wedge \mathbf{y}) \vee (\mathbf{u} \wedge \mathbf{v})$$

The *common factor theorem* states that $\mathbf{z}$ can be expressed as

$$(\mathbf{x} \wedge \mathbf{y}) \vee (\mathbf{u} \wedge \mathbf{v}) = (\mathbf{x} \wedge \mathbf{y} \wedge \mathbf{v}) \vee \mathbf{u} - (\mathbf{x} \wedge \mathbf{y} \wedge \mathbf{u}) \vee \mathbf{v} \tag{A.25}$$

A.4.5 The Intersection of Two Bivectors in a Three-Dimensional Space

As given by (A.7), in a 3-space, the exterior product of three vectors will, in any given basis, give the basis trivector, multiplied by the determinant of the components of the vectors making up the trivector. The regressive product (intersection) of a vector with an element like the basis trivector completely containing the vector gives an element congruent to itself. Thus, the intersection of the two bivectors is given by

$$\mathbf{z} = \text{Det}[\mathbf{x},\mathbf{y},\mathbf{v}]\mathbf{u} - \text{Det}[\mathbf{x},\mathbf{y},\mathbf{u}]\mathbf{v} \tag{A.26}$$

where Det[**x**,**y**,**v**] is the determinant of the components of **x**, **y**, and **v**.

A.5 Geometric Interpretations

All of the above discussion applies irrespective of any physical associations of, for example, elements and vectors, to physical objects. As noted, a metric is not necessary for the notions to apply, yielding a powerful algebra that can be used to manipulate higher order geometric objects and shapes. In this section, associations to physical objects will be established and application of the algebra to their manipulation will be demonstrated.

A.5.1 Points and Vectors

The element in a linear space is called a vector. Even though it need not be the case, the term *vector* is associated with the geometric concept of *direction*.

An element of a linear space may also be interpreted as a *point*. Of course, vectors may also be used to represent points, but only *relative to another given point* (the information for which is not in the vector). For a single vector, this other point is often the origin of the linear space. The origin of 3-D space, for example, is given by $(x, y, z) = (0, 0, 0)$, as illustrated in Figure A.2. These vectors are sometimes called *position vectors*. The reference point, however, need not be the origin and the concept of a vector can still be applied to define a point. We cannot interpret it physically in this case, however. When *both position and direction* are required in the same element, we must somehow include the information about the point of origin of the vector. A point can be represented in 3-D space by two (or more) vectors without the assumed reference to the origin, however, as illustrated in Figure A.3.

A.5.2 Sums and Differences of Points

A *point* is defined as the sum of the *origin point* and a *vector*. If o is the origin and **x** is a vector, then $o + \mathbf{x}$ is a point:

$$\mathbf{P} \triangleq o + \mathbf{x} \tag{A.27}$$

The vector **x** is the position vector of the point **P**. The *difference* of two points is a vector (not a point), since the origins cancel (remember that in Grassmann algebra,

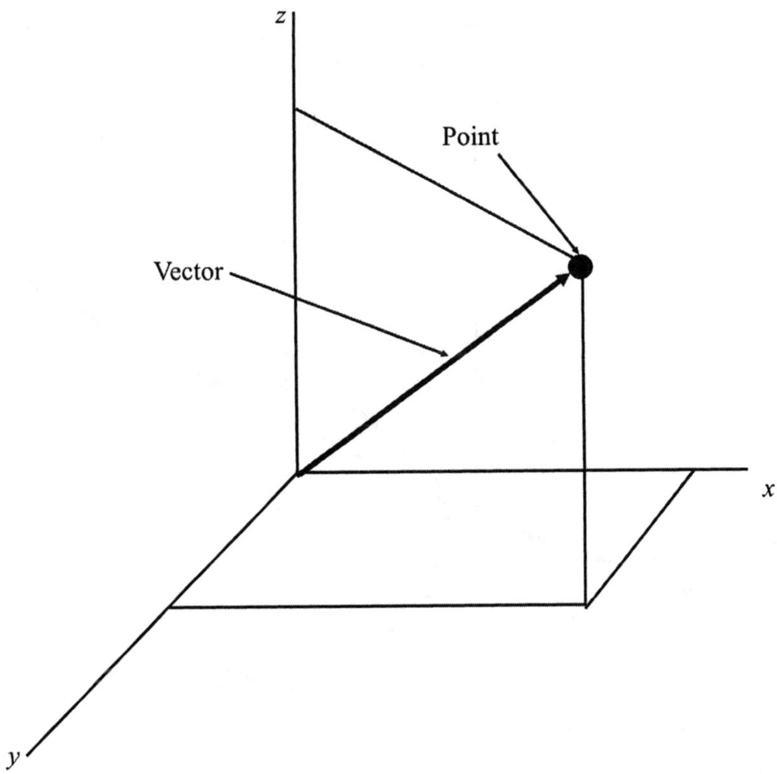

Figure A.2 Vectors and points.

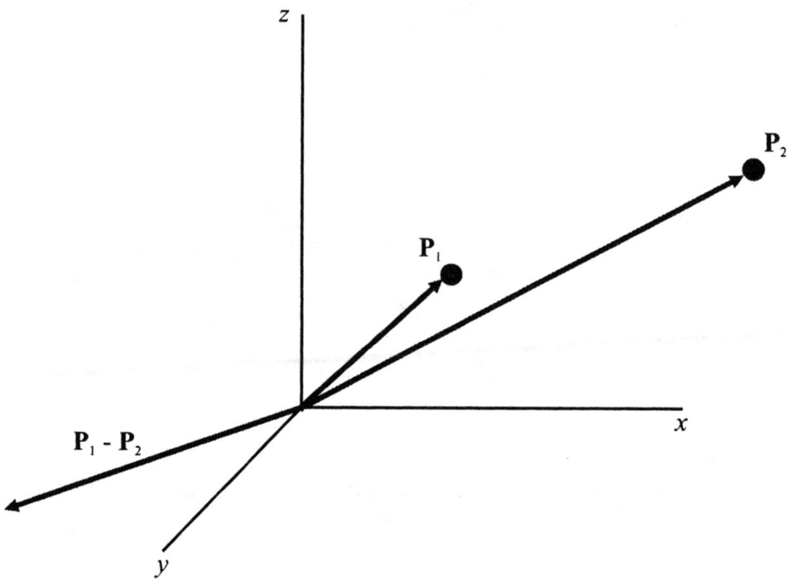

Figure A.3 Difference of two points.

the operations of addition and subtraction are the familiar ones from high school linear algebra):

$$\mathbf{P}_1 - \mathbf{P}_2 = (o + \mathbf{x}_1) - (o + \mathbf{x}_2) = \mathbf{x}_1 - \mathbf{x}_2$$

See Figure A.3.

A scalar multiple of a point is called a *weighted point*. For example, if m is a scalar, $m\mathbf{P}$ is a weighted point with weight m. The *sum* of two points with unit weights gives the point halfway between them with a weight of 2. Such a sum is illustrated in Figure A.4.

$$\mathbf{P}_1 + \mathbf{P}_2 = (o + \mathbf{x}_1) + (o + \mathbf{x}_2) = 2\left(o + \frac{\mathbf{x}_1 + \mathbf{x}_2}{2}\right)$$

A.5.3 Lines and Planes

The exterior product of a point and a vector gives a *line-bound vector*. A line is the space of a line-bound vector. That is, it consists of all the points on the line and all the vectors in the direction of the line. The reason for this is the lack of the point of origin of the vectors—all such origination points are included in the definition. Thus, all vectors in the direction of the line are included.

A line-bound vector can be represented by the line **L** *through* the point **P** in the direction of **x** and any entity *congruent* to the exterior product of **P** and **x**.

$$\mathbf{L} \equiv \mathbf{P} \wedge \mathbf{x}$$

The concept of the line-bound vector is illustrated in Figure A.5.

A line may also be represented by the exterior product of *any two points* on it. Let **Q** be any point on the line so that **Q** = **P** + a**x** is the sum of **P** and any scalar multiple of the vector. Then

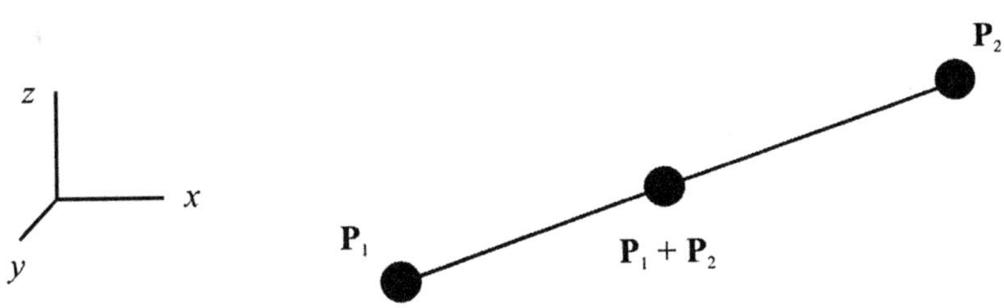

Figure A.4 Sum of two points.

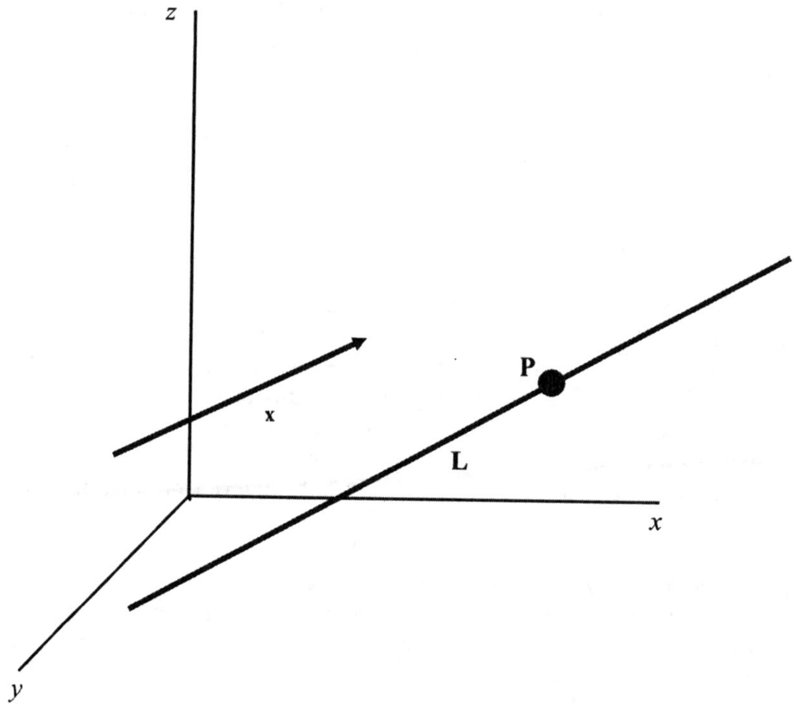

Figure A.5 Line-bound vector; a line defined by a point and a vector is the line passing through the point in the direction of the vector and any entity congruent to the exterior product of **P** and **x**.

$$\mathbf{L} \equiv \mathbf{P} \wedge \mathbf{Q} = \mathbf{P} \wedge (\mathbf{P} + a\mathbf{x}) = a\mathbf{P} \wedge \mathbf{x}$$

Therefore, as a general statement of these conditions:

$$\mathbf{L} \equiv \mathbf{P}_1 \wedge \mathbf{P}_2 \equiv \mathbf{P} \wedge \mathbf{x} \tag{A.28}$$

These concepts extend to higher dimensions in the natural way. Thus, a plane π may be represented by the exterior product of any three points on it, any two points on it together with a vector in it (not parallel to the line joining the points), or a single point on it together with a bivector in the direction of the plane:

$$\pi \equiv \mathbf{P}_1 \wedge \mathbf{P}_2 \wedge \mathbf{P}_3 \equiv \mathbf{P}_1 \wedge \mathbf{P}_2 \wedge \mathbf{x} \equiv \mathbf{P}_1 \wedge \mathbf{x} \wedge \mathbf{y} \tag{A.29}$$

Higher dimensional geometric entities are constructed from lower dimensional ones by simply computing their exterior product. A line can be obtained by taking the exterior product of a point with any point or vector *exterior* to it, as was just illustrated. Or a plane can be constructed by taking the exterior product of a line with any point or vector exterior to it.

A.5.4 Intersection of Two Lines

The regressive product is used to find intersections of geometric entities. For example, to find the intersection of two lines in a plane, which will be a point in the plane, the regressive product of the two lines is computed. The lines can be represented as

$$\mathbf{L}_1 \equiv \mathbf{P}_1 \wedge \mathbf{x}_1 = (o + \mathbf{v}_1) \wedge \mathbf{x}_1 = o \wedge \mathbf{x}_1 + \mathbf{v}_1 \wedge \mathbf{x}_1$$
$$\mathbf{L}_2 \equiv \mathbf{P}_2 \wedge \mathbf{x}_2 = (o + \mathbf{v}_2) \wedge \mathbf{x}_2 = o \wedge \mathbf{x}_2 + \mathbf{v}_2 \wedge \mathbf{x}_2$$

The point of intersection is found by

$$\mathbf{P} \equiv \mathbf{L}_1 \vee \mathbf{L}_2 = (o \wedge \mathbf{x}_1 + \mathbf{v}_1 \wedge \mathbf{x}_1) \vee (o \wedge \mathbf{x}_2 + \mathbf{v}_2 \wedge \mathbf{x}_2)$$

Four terms arise from this expression:

$$\begin{aligned}\mathbf{P} \equiv\ & (o \wedge \mathbf{x}_1) \vee (o \wedge \mathbf{x}_2) + (\mathbf{v}_1 \wedge \mathbf{x}_1) \vee (o \wedge \mathbf{x}_2) \\ & + (o \wedge \mathbf{x}_1) \vee (\mathbf{v}_2 \wedge \mathbf{x}_2) + (\mathbf{v}_1 \wedge \mathbf{x}_1) \vee + (\mathbf{v}_2 \wedge \mathbf{x}_2)\end{aligned}$$

The common factor theorem in (A.25) applied to these products generates:

$$\begin{aligned}(o\wedge\mathbf{x}_1)\vee(o\wedge\mathbf{x}_2) &= (o\wedge\mathbf{x}_1\wedge\mathbf{x}_2)\vee o-(o\wedge\mathbf{x}_1\wedge o)\vee\mathbf{x}_2\\ &= (o\wedge\mathbf{x}_1\wedge\mathbf{x}_2)\vee o\end{aligned}$$

$$\begin{aligned}(\mathbf{v}_1\wedge\mathbf{x}_1)\vee(o\wedge\mathbf{x}_2) &= (\mathbf{v}_1\wedge\mathbf{x}_1\wedge\mathbf{x}_2)\vee o-(\mathbf{v}_1\wedge\mathbf{x}_1\wedge o)\vee\mathbf{x}_2\\ &= -(\mathbf{v}_1\wedge\mathbf{x}_1\wedge o)\vee\mathbf{x}_2\end{aligned}$$

$$\begin{aligned}(o\wedge\mathbf{x}_1)\vee(\mathbf{v}_2\wedge\mathbf{x}_2) &= -(\mathbf{v}_2\wedge\mathbf{x}_2)\vee(o\wedge\mathbf{x}_1)\\ &= -(\mathbf{v}_2\wedge\mathbf{x}_2\wedge\mathbf{x}_1)\vee o+(\mathbf{v}_2\wedge\mathbf{x}_2\wedge o)\vee\mathbf{x}_1\\ &= (\mathbf{v}_2\wedge\mathbf{x}_2\wedge o)\vee\mathbf{x}_1\end{aligned}$$

$$\begin{aligned}(\mathbf{v}_1\wedge\mathbf{x}_1)\vee(\mathbf{v}_2\wedge\mathbf{x}_2) &= (\mathbf{v}_1\wedge\mathbf{x}_1\wedge\mathbf{x}_2)\vee\mathbf{v}_2-(\mathbf{v}_1\wedge\mathbf{x}_1\wedge\mathbf{v}_2)\vee\mathbf{x}_2\\ &= 0\end{aligned}$$

These simplifications arise because the following facts about vectors and exterior products were applied: (1) the term $(o\wedge\mathbf{x}_1\wedge o)$ is zero because of the exterior product of repeated factors; and (2) the four terms involving the exterior product of three vectors, for example $(\mathbf{v}_1\wedge\mathbf{x}_1\wedge\mathbf{x}_2)$, are also zero, since any three vectors in a two-dimensional vector space must be dependent. Thus,

$$\mathbf{P}\triangleq(o\wedge\mathbf{x}_1\wedge\mathbf{x}_2)\vee o+(o\wedge\mathbf{v}_2\wedge\mathbf{x}_2)\vee\mathbf{x}_1-(o\wedge\mathbf{v}_1\wedge\mathbf{x}_1)\vee\mathbf{x}_2$$

Given a basis with basis vectors given by $\mathbf{e}_1$ and $\mathbf{e}_2$, this expression can be represented as

$$\mathbf{P}=o+\frac{\mathrm{Det}[\mathbf{v}_2,\mathbf{x}_2]}{\mathrm{Det}[\mathbf{x}_1,\mathbf{x}_2]}\mathbf{x}_1-\frac{\mathrm{Det}[\mathbf{v}_1,\mathbf{x}_1]}{\mathrm{Det}[\mathbf{x}_1,\mathbf{x}_2]}\mathbf{x}_2$$

where the coefficients of the vectors are the coefficients in the given basis.

To verify that **P** does indeed lie on both the lines $\mathbf{L}_1$ and $\mathbf{L}_2$, we only need to carry out the straightforward verification that the products $\mathbf{P}\wedge\mathbf{L}_1$ and $\mathbf{P}\wedge\mathbf{L}_2$ are both zero. Thus,

$$
\begin{aligned}
\mathbf{P} \wedge \mathbf{L}_1 &= (o \wedge \mathbf{x}_1 + \mathbf{v}_1 \mathbf{x}_1) \wedge \left[o + \frac{\mathrm{Det}(\mathbf{v}_2, \mathbf{x}_2)}{\mathrm{Det}(\mathbf{x}_1, \mathbf{x}_2)} \mathbf{x}_1 + \frac{\mathrm{Det}(\mathbf{v}_1, \mathbf{x}_1)}{\mathrm{Det}(\mathbf{x}_1, \mathbf{x}_2)} \mathbf{x}_2 \right] \\
&= o \wedge \mathbf{x}_1 \wedge o + \frac{\mathrm{Det}(\mathbf{v}_2, \mathbf{x}_2)}{\mathrm{Det}(\mathbf{x}_1, \mathbf{x}_2)} o \wedge \mathbf{x}_1 \wedge \mathbf{x}_1 \\
&\quad + \frac{\mathrm{Det}(\mathbf{v}_1, \mathbf{x}_1)}{\mathrm{Det}(\mathbf{x}_1, \mathbf{x}_2)} o \wedge \mathbf{x}_1 \wedge \mathbf{x}_2 + \mathbf{v}_1 \wedge \mathbf{x}_1 \wedge o \\
&\quad + \frac{\mathrm{Det}(\mathbf{v}_1, \mathbf{x}_1)}{\mathrm{Det}(\mathbf{x}_1, \mathbf{x}_2)} \mathbf{v}_1 \wedge \mathbf{x}_1 \wedge \mathbf{x}_1 + \frac{\mathrm{Det}(\mathbf{v}_2, \mathbf{x}_2)}{\mathrm{Det}(\mathbf{x}_1, \mathbf{x}_2)} \mathbf{v}_1 \wedge \mathbf{x}_1 \wedge \mathbf{x}_2
\end{aligned}
$$

Now

$$
o \wedge \mathbf{x}_1 \wedge o = 0 \qquad \frac{\mathrm{Det}(\mathbf{v}_2, \mathbf{x}_2)}{\mathrm{Det}(\mathbf{x}_1, \mathbf{x}_2)} o \wedge \mathbf{x}_1 \wedge \mathbf{x}_1 = 0 \qquad \frac{\mathrm{Det}(\mathbf{v}_2, \mathbf{x}_2)}{\mathrm{Det}(\mathbf{x}_1, \mathbf{x}_2)} \mathbf{v}_1 \wedge \mathbf{x}_1 \wedge \mathbf{x}_1 = 0
$$

because of the exterior product of like elements, leaving

$$
\begin{aligned}
&\frac{\mathrm{Det}(\mathbf{v}_1, \mathbf{x}_1)}{\mathrm{Det}(\mathbf{x}_1, \mathbf{x}_2)} o \wedge \mathbf{x}_1 \wedge \mathbf{x}_2 + \mathbf{v}_1 \wedge \mathbf{x}_1 \wedge o + \frac{\mathrm{Det}(\mathbf{v}_1, \mathbf{x}_1)}{\mathrm{Det}(\mathbf{x}_1, \mathbf{x}_2)} \mathbf{v}_1 \wedge \mathbf{x}_1 \wedge \mathbf{x}_2 \\
&\quad = \frac{\mathrm{Det}(\mathbf{v}_1, \mathbf{x}_1)}{\mathrm{Det}(\mathbf{x}_1, \mathbf{x}_2)} [o + \mathbf{v}_1] \wedge \mathbf{x}_1 \wedge \mathbf{x}_2 + \mathbf{v}_1 \wedge \mathbf{x}_1 \wedge o
\end{aligned} \tag{A.30}
$$

The first term on the right in (A.30) is zero because three elements in a 2-space are, of necessity, dependent and therefore their exterior product is zero. Thus

$$
\mathbf{P} \wedge \mathbf{L}_1 = \mathbf{v}_1 \wedge \mathbf{x}_1 \wedge o = 0 \tag{A.31}
$$

Therefore **P** lies on $\mathbf{L}_1$. It can be similarly shown that **P** lies on $\mathbf{L}_2$. Therefore **P** is the intersection of $\mathbf{L}_1$ and $\mathbf{L}_2$.

This development is more complicated than for the simple linear algebra solution to the same problem. However, this can be generalized to any dimension, and therefore any geometrical object, not just lines and points. The intersection of arbitrary geometrical shapes can be found. Furthermore a metric was not required to obtain the result.

A.6 The Complement

A.6.1 The Complement as a Correspondence Between Spaces

Given a linear space of dimension N with basis $\mathbf{e}_1$, $\mathbf{e}_2$, …, $\mathbf{e}_N$, the set of all the different m-element products of these basis elements forms the basis of another linear space of dimension $\binom{N}{m}$. The antisymmetric nature of the exterior product means that there are just as many basis elements in the linear space of $(n - m)$-elements as there are in the linear space of m-elements. There is a correspondence between the elements in the space of dimension N and that of dimension $N - m$. The $(N - m)$-element is called the *complement* of the m-element.

A.6.2 The Euclidean Complement

In a three-dimensional linear space with basis $\mathbf{e}_1$, $\mathbf{e}_2$, and $\mathbf{e}_3$, the *Euclidean complement* of each of the basis elements is defined as the basis 2-element whose exterior product with the basis element gives the basis 3-element $\mathbf{e}_1 \wedge \mathbf{e}_2 \wedge \mathbf{e}_3$. Thus,

$$\overline{\mathbf{e}}_1 \triangleq \mathbf{e}_2 \wedge \mathbf{e}_3 \Rightarrow \overline{\mathbf{e}}_1 \wedge \mathbf{e}_1 = \mathbf{e}_1 \wedge \mathbf{e}_2 \wedge \mathbf{e}_3 \tag{A.32}$$

$$\overline{\mathbf{e}}_2 \triangleq \mathbf{e}_3 \wedge \mathbf{e}_1 \Rightarrow \overline{\mathbf{e}}_2 \wedge \mathbf{e}_2 = \mathbf{e}_1 \wedge \mathbf{e}_2 \wedge \mathbf{e}_3 \tag{A.33}$$

$$\overline{\mathbf{e}}_3 \triangleq \mathbf{e}_1 \wedge \mathbf{e}_2 \Rightarrow \overline{\mathbf{e}}_3 \wedge \mathbf{e}_3 = \mathbf{e}_1 \wedge \mathbf{e}_2 \wedge \mathbf{e}_3 \tag{A.34}$$

The Euclidean complement of a general 1-element, $\mathbf{x} = a\mathbf{e}_1 + b\mathbf{e}_2 + c\mathbf{e}_3$, can now be defined by extending the definition on the basis elements by linearity,

$$\overline{\mathbf{x}} = \overline{a\mathbf{e}_1 + b\mathbf{e}_2 + c\mathbf{e}_3} \triangleq a\overline{\mathbf{e}}_1 + b\overline{\mathbf{e}}_2 + c\overline{\mathbf{e}}_3 = a(\mathbf{e}_2 \wedge \mathbf{e}_3) + b(\mathbf{e}_3 \wedge \mathbf{e}_1) + c(\mathbf{e}_1 \wedge \mathbf{e}_2)$$

The product $\mathbf{x} \wedge \overline{\mathbf{x}}$ expands to

$$\begin{aligned}\mathbf{x} \wedge \overline{\mathbf{x}} &= (a\mathbf{e}_1 + b\mathbf{e}_2 + c\mathbf{e}_3) \wedge (a\mathbf{e}_2 \wedge \mathbf{e}_3 + b\mathbf{e}_3 \wedge \mathbf{e}_1 + c\mathbf{e}_1 \wedge \mathbf{e}_2) \\ &= (a^2 + b^2 + c^2)\mathbf{e}_1 \wedge \mathbf{e}_2 \wedge \mathbf{e}_3\end{aligned}$$

The complements of the basis 2-elements is defined just as those for 1-elements, such that the exterior product of a basis 2-element with its complement is equal to the basis 3-element. The complement of a 2-element in 3-space is therefore a 1-element

$$\overline{\mathbf{e}_2 \wedge \mathbf{e}_3} = \mathbf{e}_1 \Rightarrow \mathbf{e}_2 \wedge \mathbf{e}_3 \wedge \overline{\mathbf{e}_2 \wedge \mathbf{e}_3} = \mathbf{e}_2 \wedge \mathbf{e}_3 \wedge \mathbf{e}_1 = \mathbf{e}_1 \wedge \mathbf{e}_2 \wedge \mathbf{e}_3$$
$$\overline{\mathbf{e}_3 \wedge \mathbf{e}_1} = \mathbf{e}_2 \Rightarrow \mathbf{e}_3 \wedge \mathbf{e}_1 \wedge \overline{\mathbf{e}_3 \wedge \mathbf{e}_1} = \mathbf{e}_3 \wedge \mathbf{e}_1 \wedge \mathbf{e}_2 = \mathbf{e}_1 \wedge \mathbf{e}_2 \wedge \mathbf{e}_3$$
$$\overline{\mathbf{e}_1 \wedge \mathbf{e}_2} = \mathbf{e}_3 \Rightarrow \mathbf{e}_1 \wedge \mathbf{e}_2 \wedge \overline{\mathbf{e}_1 \wedge \mathbf{e}_2} = \mathbf{e}_1 \wedge \mathbf{e}_2 \wedge \mathbf{e}_3$$

In addition,

$$\overline{\mathbf{1}} = \mathbf{e}_1 \wedge \mathbf{e}_2 \wedge \mathbf{e}_3$$
$$\overline{\mathbf{e}_1 \wedge \mathbf{e}_2 \wedge \mathbf{e}_3} = \mathbf{1}$$

A.6.3 The Complement of a Complement

The complement of the complement of **x** is just **x** itself.

$$\mathbf{x} = a\mathbf{e}_1 \wedge b\mathbf{e}_2 \wedge c\mathbf{e}_3$$
$$\overline{\mathbf{x}} = \overline{a\mathbf{e}_1 \wedge b\mathbf{e}_2 \wedge c\mathbf{e}_3} = a\overline{\mathbf{e}}_1 \wedge b\overline{\mathbf{e}}_2 \wedge c\overline{\mathbf{e}}_3 = a\mathbf{e}_2 \wedge \mathbf{e}_3 + b\mathbf{e}_3 \wedge \mathbf{e}_1 + c\mathbf{e}_1 \wedge \mathbf{e}_2$$
$$\overline{\overline{\mathbf{x}}} = \overline{a\mathbf{e}_2 \wedge \mathbf{e}_3 + b\mathbf{e}_3 \wedge \mathbf{e}_1 + c\mathbf{e}_1 \wedge \mathbf{e}_2}$$
$$= a\overline{\mathbf{e}_2 \wedge \mathbf{e}_3} + b\overline{\mathbf{e}_3 \wedge \mathbf{e}_1} + c\overline{\mathbf{e}_1 \wedge \mathbf{e}_2} = a\mathbf{e}_1 + b\mathbf{e}_2 + c\mathbf{e}_3$$

The complement of *any* element is the element itself, apart from a possible sign:

$$\overline{\overline{\underset{m}{\boldsymbol{\alpha}}}} = (-1)^{m(N-m)} \underset{m}{\boldsymbol{\alpha}} \qquad \text{(A.35)}$$

A.6.4 The Complement Axiom

The complement axiom states that the Euclidean complement of the exterior product of two elements is equal to the regressive product of their complements:

$$\overline{\underset{m}{\boldsymbol{\alpha}} \wedge \underset{k}{\boldsymbol{\beta}}} = \overline{\underset{m}{\boldsymbol{\alpha}}} \vee \overline{\underset{k}{\boldsymbol{\beta}}} \qquad \text{(A.36)}$$

This expression is similar to de Morgan's law in Boolean algebra.

A.7 The Interior Product

The *interior product* of an element $\underset{m}{\boldsymbol{\alpha}}$ with an element $\underset{k}{\boldsymbol{\beta}}$ is denoted by $\underset{m}{\boldsymbol{\alpha}} \odot \underset{k}{\boldsymbol{\beta}}$ and *defined* to be the *regressive* product of $\underset{m}{\boldsymbol{\alpha}}$ with the complement of $\underset{k}{\boldsymbol{\beta}}$:

$$\underset{m}{\boldsymbol{\alpha}} \odot \underset{k}{\boldsymbol{\beta}} \triangleq \underset{m}{\boldsymbol{\alpha}} \vee \overline{\underset{k}{\boldsymbol{\beta}}} \tag{A.37}$$

The grade of an interior product $\underset{m}{\boldsymbol{\alpha}} \odot \underset{k}{\boldsymbol{\beta}}$ is $m + (N - k) - N = m - k$.

The interior product is a generalization of the inner product of scalars and vectors.

A.7.1 Inner Products and Scalar Products

The interior product of two elements $\underset{m}{\boldsymbol{\alpha}}$ and $\underset{m}{\boldsymbol{\beta}}$ *of the same grade* is called their *inner product*. Since the grade of an interior product is the difference of the grades of its factors, an inner product is always of grade zero, hence scalar.

When the two factors of the product are grade 1, the inner product is called a *scalar product*, which is the definition used in everyday life.

The inner product is symmetric, that is,

$$\underset{m}{\boldsymbol{\alpha}} \odot \underset{m}{\boldsymbol{\beta}} = \underset{m}{\boldsymbol{\beta}} \odot \underset{m}{\boldsymbol{\alpha}} \tag{A.38}$$

A.7.2 Calculating Interior Products

Let the linear space under consideration be a 3-space with Euclidean metric and basis elements e_1, e_2, and e_3. The scalar products $\mathbf{e}_1 \odot \mathbf{e}_1$ and $\mathbf{e}_1 \odot \mathbf{e}_2$ are

$$\mathbf{e}_1 \odot \mathbf{e}_1 = \mathbf{e}_1 \vee \overline{\mathbf{e}}_1 = \mathbf{e}_1 \vee (\mathbf{e}_2 \wedge \mathbf{e}_3) = (\mathbf{e}_1 \wedge \mathbf{e}_2 \wedge \mathbf{e}_3) \vee \mathbf{1} = \overline{\mathbf{1}} \vee \mathbf{1} = 1$$
$$\mathbf{e}_1 \odot \mathbf{e}_2 = \mathbf{e}_1 \vee \overline{\mathbf{e}}_2 = \mathbf{e}_1 \vee (\mathbf{e}_3 \wedge \mathbf{e}_1) = (\mathbf{e}_1 \wedge \mathbf{e}_3 \wedge \mathbf{e}_1) \vee \mathbf{1} = \mathbf{0} \vee \mathbf{1} = 0$$

That is, the scalar product of a basis element with itself is one while the scalar product of two different basis functions is zero.

Since identical basis 2-elements are the same element, their inner product is one:

$$(\mathbf{e}_1 \wedge \mathbf{e}_2) \odot (\mathbf{e}_1 \wedge \mathbf{e}_2) = (\mathbf{e}_1 \wedge \mathbf{e}_2) \vee \overline{(\mathbf{e}_1 \wedge \mathbf{e}_2)}$$
$$= (\mathbf{e}_1 \wedge \mathbf{e}_2) \vee \mathbf{e}_3 = (\mathbf{e}_1 \wedge \mathbf{e}_2 \wedge \mathbf{e}_3) \vee \mathbf{1} = \overline{\mathbf{1}} \vee \mathbf{1} = 1$$

Likewise, the inner product of nonidentical basis 2-elements is zero:

$$(\mathbf{e}_1 \wedge \mathbf{e}_2) \odot (\mathbf{e}_2 \wedge \mathbf{e}_3) = (\mathbf{e}_1 \wedge \mathbf{e}_2) \vee \overline{(\mathbf{e}_2 \wedge \mathbf{e}_3)}$$
$$= (\mathbf{e}_1 \wedge \mathbf{e}_2) \vee \mathbf{e}_1 = (\mathbf{e}_1 \wedge \mathbf{e}_2 \wedge \mathbf{e}_1) \vee \mathbf{1} = 0 \vee \mathbf{1} = 0$$

If a basis 2-element contains a given basis 1-element, then their interior product is not zero:

$$(\mathbf{e}_1 \wedge \mathbf{e}_2) \odot \mathbf{e}_1 = (\mathbf{e}_1 \wedge \mathbf{e}_2) \vee \overline{\mathbf{e}_1}$$
$$= (\mathbf{e}_1 \wedge \mathbf{e}_2) \vee (\mathbf{e}_2 \wedge \mathbf{e}_3) = (\mathbf{e}_1 \wedge \mathbf{e}_2 \wedge \mathbf{e}_3) \vee \mathbf{e}_2 = \overline{\mathbf{1}} \vee \mathbf{e}_2 = \mathbf{e}_2$$

If a basis 2-element does *not* contain a given basis 1-element, then their interior product is zero:

$$(\mathbf{e}_1 \wedge \mathbf{e}_2) \odot \mathbf{e}_3 = (\mathbf{e}_1 \wedge \mathbf{e}_2) \vee \overline{\mathbf{e}_3}$$
$$= (\mathbf{e}_1 \wedge \mathbf{e}_2) \vee (\mathbf{e}_1 \wedge \mathbf{e}_2) = 0$$

A.7.3 Expanding Interior Products

Theorem: Interior Common Factor Theorem.
If $\underset{m}{\alpha}$ is a simple element,

$$\underset{m}{\alpha} \odot \underset{k}{\beta} = \sum_{i=1}^{\nu} \left(\underset{k}{\alpha_i} \odot \underset{k}{\beta_i} \right) \underset{m-k}{\alpha_i} \qquad k \le m, \quad \nu = \binom{m}{k}$$
$$\underset{m}{\alpha} = \underset{k}{\alpha_1} \wedge \underset{m-k}{\alpha_1} = \underset{k}{\alpha_2} \wedge \underset{m-k}{\alpha_2} = \cdots = \underset{k}{\alpha_\nu} \wedge \underset{m-k}{\alpha_\nu}$$

In these expressions $\underset{k}{\beta}$ need not be simple. Thus, the interior product of a simple element $\underset{m}{\alpha}$ with another element of equal or lower grade $\underset{m}{\beta}$ may be expressed as a linear combination of the ν essentially different factors $\underset{m-k}{\alpha}$ (of grade $m - k$) of the simple element of higher degree.

A.7.4 The Interior Product of a Bivector and a Vector

Given that $\mathbf{x}$ is a vector and that $\mathbf{x}_1 \wedge \mathbf{x}_2$ is a simple bivector. The interior product of $\mathbf{x}_1 \wedge \mathbf{x}_2$ with $\mathbf{x}$ is the vector $(\mathbf{x}_1 \wedge \mathbf{x}_2) \odot \mathbf{x}$. By the Interior Common Factor Theorem, or the formula above derived from it, this can be expanded to give

$$(\mathbf{x}_1 \wedge \mathbf{x}_2) \odot \mathbf{x} = (\mathbf{x} \odot \mathbf{x}_1)\mathbf{x}_2 - (\mathbf{x} \odot \mathbf{x}_2)\mathbf{x}_1$$

which is a linear combination of $\mathbf{x}_1$ and $\mathbf{x}_2$. Thus, it is clearly contained in the bivector $\mathbf{x}_1 \wedge \mathbf{x}_2$. Taking the result's scalar product with $\mathbf{x}$,

$$((\mathbf{x} \odot \mathbf{x}_1)\mathbf{x}_2 - (\mathbf{x} \odot \mathbf{x}_2)\mathbf{x}_1) \odot \mathbf{x} = (\mathbf{x} \odot \mathbf{x}_1)(\mathbf{x}_2 \odot \mathbf{x}) - (\mathbf{x} \odot \mathbf{x}_2)(\mathbf{x}_1 \odot \mathbf{x}) = \mathbf{0}$$

shows that the result is orthogonal to $\mathbf{x}$.

These concepts are not restricted to two dimensions and may readily be extended to geometric entities of arbitrary dimension.

A.8 Concluding Remarks

The basics of an algebra discovered by Hermann Grassmann in the nineteenth century were presented in this appendix. The application of concern to EW target geolocation was presented in Chapter 3 (Section 3.3). That is by no means the only application of this algebra however.

Grassmann algebra subsumes the linear algebra that most learn about in high school. By making the appropriate associations of exterior and interior products with everyday products the mapping becomes clear.

References

[1] Browne, J., *Grassmann Algebra, Exploring Applications of Extended Vector Algebra with Mathematica,* incomplete draft, www.ses.swin.edu.au/home/browne/grassmannalgebra/book/.

[2] Armstrong, H. L., "On an Alternative Definition of the Vector Product in *n*-Dimensional Vector Analysis," *Matrix and Tensor Quarterly*, Vol. IX, No. 4, 1959, pp. 107–110.

[3] Barton, H., "A Modern Presentation of Grassmann's Tensor Analysis," *American Journal of Mathematics*, Vol. XLIX, 1927, pp. 598–614.

[4] Bowen, R. M., and C.-C. Wang, *Introduction to Vectors and Tensors,* New York: Plenum Press, 1976.

[5] Burali-Forti, C., *Introduction à la Géométrie Différentielle suivant la Méthode de H. Grassmann,* Paris: Gauthier-Villars, 1897.

[6] Burali-Forti, C., and R. Marcolongo, *Éléments de Calcul Vectoriel,* Paris: Hermann, 1910.

[7] Carvallo, M. E., "La Méthode de Grassmann," *Nouvelles Annales de Mathématiques*, serie 3 Vol. XI, 1892, pp. 8–37.

[8] Chevalley, C., *The Construction and Study of Certain Important Algebras,* Tokyo: Mathematical Society of Japan, 1955.

[9] Clifford, W. K., "Applications of Grassmann's Extensive Algebra," *American Journal of Mathematics Pure and Applied*, Vol. I, 1878, pp. 350–358.

[10] Clifford, W. K., *Mathematical Papers,* 1882. Reprinted by Chelsea Publishing Co. New York, 1968.

[11] Coffin, J. G., *Vector Analysis,* New York: Wiley, 1909.

[12] Collins, J. V., "An Elementary Exposition of Grassmann's *Ausdehnungslehre* or Theory of Extension," *American Mathematical Monthly*, Vol. 6 (1899), pp. 193–198, 261–266, 297–301; Vol. 7 (1900), pp. 31–35, 163–166, 181–187, 207–214, 253–258.

[13] Coolidge, J. L., "Grassmann's Calculus of Extension" in *A History of Geometrical Methods,* Oxford: Oxford University Press, 1940, pp. 252–257.

[14] Crowe, M. J., *A History of Vector Analysis,* Notre Dame: Notre Dame, 1967.

[15] Dibag, I., "Factorization in Exterior Algebras," *Journal of Algebra,* New York: Academic Press, Vol. 30, 1974, pp. 259–262.

[16] Gibbs, J. W., *The Collected Works of J. Willard Gibbs Ph.D. L.L.D.,* New York: Longmans, 1928.

[17] Grassmann, H. G., *Gesammelte mathematische und physikalische Werke,* F. Engel, Ed., Teubner: Leipzig, Vol. 1 (1896), Vol. 2 (1902, 1904), and Vol. 3 (1911).

[18] Kannenberg., L. C., *A New Branch of Mathematics: The Ausdehnungslehre of 1844 and Other Works,* Chicago: Open Court, 1995.

[19] Whitehead, A. N., *A Treatise on Universal Algebra,* Cambridge: Cambridge University Press, 1898.

[20] Hyde, E. W., *The Directional Calculus,* Boston: Ginn & Co., 1890.

[21] Zaddach, A., *Grassmann's Algebra in der Geometrie*, Zurich: BI-Wissenschaftsverlag, 1994.

List of Acronyms

AM	amplitude modulation
AOA	angle of arrival
AOI	area of interest
AWGN	additive white Gaussian noise
CAF	cross ambiguity function
CEP	circular error probable
CRLB	Cramer-Rao lower bound
DD	differential Doppler
det	determinate
DF	direction finding
DME	distance measuring equipment
DOP	dilution of precision
DPD	discrete probability density
EEP	elliptical error probable
EKF	extended Kalman filter
EM	electromagnetic
EW	electronic warfare
FCC	Federal Communications Commission
FDOA	frequency difference of arrival
FM	frequency modulation
FOA	frequency of arrival
GCM	generalized correlation method
GDOP	geometric dilution of precision
GMT	Greenwich mean time
GPS	global positioning system
GT	Greenwich time
HF	high frequency

Hz	Hertz
IEE	Institute of Electrical Engineers
IEEE	Institute of Electrical and Electronic Engineers
i.i.d.	independent and identically distributed
KF	Kalman filter
LMMSE	linear minimum mean-squared error
LOCA	location on the conic axis
LOB	line of bearing
LOP	line of position
LS	least-squared
LSE	least-squared error
LUF	lowest usable frequency
mHz	millihertz
MHz	megahertz
ML	maximum likelihood
MLE	maximum likelihood estimation
MMSE	minimum mean-squared error
MUF	maximum usable frequency
NVIS	near vertical incidence sky wave
PF	position fix
PM	phase modulation
PX	plane intersection
RD	range difference
SDOA	scale difference of arrival
SI	spherical interpolation
SNR	signal-to-noise ratio
SSL	single-site location
SVD	singular value decomposition
SX	spherical intersection
TDOA	time difference of arrival
TLS	total least-squared
TLSE	total least squared error
TMA	target motion analysis

TOA	time of arrival
UAV	unattended aerial vehicle
μsec	microsecond
UT	universal time
VHF	very high frequency
WLSE	weighted least-squared error

About the Author

Richard A. Poisel received a B.S. in electrical engineering from the Milwaukee School of Engineering in 1969 and an M.S. in the same discipline from Purdue University in 1971. He spent three years in the military service from 1971 to 1973. After his service he attended the University of Wisconsin, where he received a Ph.D. in electrical and computer engineering in 1977. From 1977 to 2004 he was with the same government organization, which has had several different names and is currently known as the U.S. Army Research, Development, and Engineering Command, Intelligence and Information Warfare Laboratory. During the 1993–1994 academic year, Dr. Poisel attended the MIT Sloan School of Management as a Sloan Fellow, receiving an M.B.A. Initially a research engineer, Dr. Poisel eventually rose to the role of the director of the laboratory on an acting basis from 1997 to 1999. He was appointed chief scientist in 1999 and was relocated to the Army's Intelligence Center at Ft. Huachuca, Arizona, where he served as a technical advisor to the command group. He is currently employed by Raytheon Missile Systems as a senior engineering fellow.

Index

The Artech House Information Warfare Library

Electronic Intelligence: The Analysis of Radar Signals, Second Edition, Richard G. Wiley

Electronic Warfare for the Digitized Battlefield, Michael R. Frater and Michael Ryan

Electronic Warfare in the Information Age, D. Curtis Schleher

Electronic Warfare Target Location Methods, Richard A. Poisel

EW 101: A First Course in Electronic Warfare, David Adamy

Information Warfare Principles and Operations, Edward Waltz

Introduction to Communication Electronic Warfare Systems, Richard A. Poisel

Knowledge Management in the Intelligence Enterprise, Edward Waltz

Mathematical Techniques in Multisensor Data Fusion, Second Edition, David L. Hall and Sonya A. H. McMullen

Modern Communications Jamming Principles and Techniques, Richard A. Poisel

Principles of Data Fusion Automation, Richard T. Antony

Tactical Communications for the Digitized Battlefield, Michael Ryan and Michael R. Frater

Target Acquisition in Communication Electronic Warfare Systems, Richard A. Poisel